6-18-57 L.G. 72

.50
.35
.25
.15

D1297315

THE EDUCATION OF T. C. MITS

THE
EDUCATION
OF
T. C. MITS

Drawings by
HUGH GRAY LIEBER

Words by
LILLIAN R. LIEBER

NEW YORK

W. W. Norton & Company, Inc.

PRINTED IN THE UNITED STATES OF AMERICA
FOR THE PUBLISHERS BY THE NORWOOD PRESS

PREFACE

This is not intended to be
free verse.
Writing each phrase on a separate line
facilitates rapid reading,
and everyone
is in a hurry
nowadays.

CONTENTS

THE MORAL

CONTENTS

INTRODUCING THE HERO—T. C. MITS

This introduces the Hero:

T.	C.	M	I	T	S
h	e	a	n	h	t
e	l	n		e	r
	e				e
	b				e
	r				t
	a				
	t				
	e				
	d				

T. C. is born and gets
an education of some kind—
perhaps college,
perhaps "the school of hard knocks."
In any case
he tries to figure out
how best to "get along."
He picks up a lot of
contradictory information:

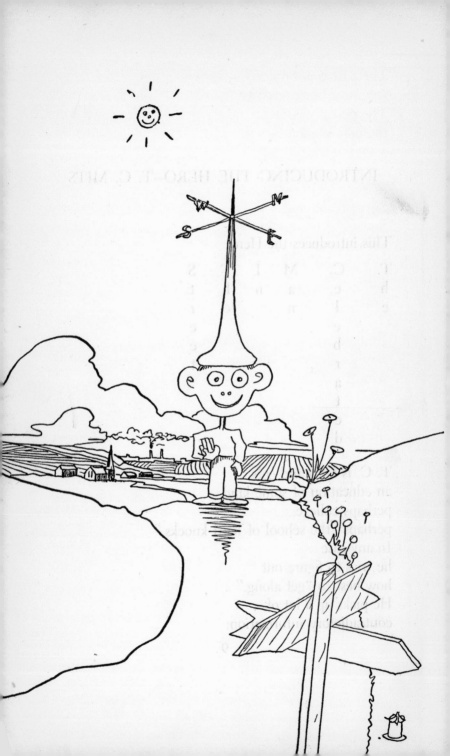

"The past is antiquated,
   you must be progressive."
"The past is wonderful,
   the new-fangled fads are
   a sign of decadence."

"Science will save us from
   Superstition and Fraud."
"Science is the greatest menace
   yet invented by man."

"Fifty million people can't be wrong."
"Some races are always wrong."

"Be practical, learn a vocation,
   don't waste your time on
   Mathematics and Art."

"Why be a narrow, practical farmer
   all your life,
   get out and learn some theory,
   and find out how
   to do things in a better way."

And so on and so on.

He is naturally confused by all this,
and very much hemmed in.
He becomes not only
Mits in name,

11

but has mits on his fingers
and mits on his toes,
and is generally "mitsified"
in the brain.

This book is an attempt
to get a bird's-eye view
of T. C.'s predicament,
and to look for
a possible egress.

To do this
VIVIDLY,
we use pictures whenever possible.
And
to do it
CLEARLY,
we use the clearest language
man has invented:
Mathematics.

Oh, we know you do not like
Mathematics,
but
we promise not to
use it as an instrument of torture,
but to show
what bearing it may have
on the contradictory advice
mentioned above,
as well as on such things as:

12

Democracy
Freedom and License
Pride and Prejudice
Success
Isolationism
Preparedness
Tradition
Progress
Idealism
Common Sense
Human Nature
War
Self-reliance
Humility
Tolerance
Provincialism
Anarchy
Loyalty
Abstract Art
and so on.

Now and then we shall point out
a "Moral."
But please do not think
we are being didactic
and preaching to the reader:

the fact is that we are really
talking to ourselves,
for we, along with millions of others,
are T. C. himself.

PART I

THE OLD

I. FIFTY MILLION PEOPLE CAN BE WRONG

Let us begin with
a very simple question:
suppose you had the choice of
the following two jobs:

Job 1: Starting with an
annual salary of $1000,
and a $200 increase every year.

Job 2: Starting with a
semiannual salary of $500,
and an increase of
$50 every 6 months.

In all other respects,
the two jobs are exactly alike.

Which is the better offer
(after the first year)?
Think carefully and
decide on your answer
BEFORE TURNING THIS PAGE.

Did you say Job 1 is better?
And did you reason as follows?
Since Job 2 has an increase
of $50 every 6 months,
it must have an annual increase of $100
and therefore it is not as good
as Job 1 which has
an annual increase of $200.

Well, you are wrong!
For, examine carefully
the earnings written out below:

			1st half of year	2nd half of year	total for the year
1st year	{	Job 1	$500	$500	$1000
		Job 2	500	550	1050
2nd year	{	Job 1	600	600	1200
		Job 2	600	650	1250
3rd year	{	Job 1	700	700	1400
		Job 2	700	750	1450
4th year	{	Job 1	800	800	1600
		Job 2	800	850	1650

etc., etc., etc.

Note that:
(1) Job 1 pays $200 more each year
    than it did the previous year.
(2) Job 2 pays $50 more every
    half-year than it did during
    the previous half-year.

18

All this is in accordance with
the promises originally made,
and yet
Job 2 brings in $50 more every year
than Job 1 does.
And you can easily see that
this will continue to be true
no matter what number of years
is considered.

You are probably surprised.
But don't be discouraged,
for you are in plenty of
good company.
Try it on your friends,
and you will find that,
unless they have heard it before,
they will probably make
the same mistake that you made.
Fifty million people CAN be wrong!
And this is entirely normal.
But please do not come to
the conclusion that
Democracy is no good!
For fifty million people
do not HAVE to be wrong!
They may be wrong when
they are too hasty and
jump at conclusions,
as you saw in the problem above.

So do not make a similar mistake
again
by coming to hasty conclusions about
Democracy.
We are coming back to Democracy
later.

In the meantime, please remember that
you can fool
"ALL of the people SOME of the time
  but NOT ALL the people ALL the time."

And since you are one of the people
yourself,
and don't want to be fooled
if you can help it,
you must be prepared to think straight.
And, incidentally,
don't fool yourself either
by thinking that this can be done
without any effort at all on your part.
Perhaps this little book will help
to smooth the road for you.

The Moral: Don't be a
            Conclusion-Jumper.

II. DON'T HIT THE CEILING

Let us try another one,
and this time give it a little more thought:
Suppose you had a paper napkin,
say about three-thousandths (.003)
of an inch thick.
Now lay another similar napkin on top of it;
the two will of course be twice as thick as one,
$$.003 \times 2 = .006,$$
or six-thousandths of an inch thick.
Now put two more napkins on top of that
making 4 in all,
which are $.003 \times 4 = .012$,
or twelve-thousandths of an inch thick.
Continue this process,
each time doubling the number of napkins,
thus:
the first time you had 1 napkin,
the second time you had 2,
the third time, 4,
the fourth time, 8,
the fifth time, 16,
and so on,

$2 \times 2 \times 2 \times 2 \times 2 \times 2 \times 2 \times 2 \times 2 \times 2 \times 2^{\sim}$

doubling the number each time,
as we said before.
Now continue this 32 times.
The question is:
HOW HIGH WILL THE PILE OF NAPKINS BE?
Do you think it will be 1 foot high?
Or will it be as high as
a normal room, from floor to ceiling?
Or as high as the Empire State Building?
Or what?
The correct answer is not necessarily
any of these.
What do YOU think?
Decide BEFORE you turn this page.

Let us again make out a table,
showing clearly what was done:

		NO. OF NAPKINS	THICKNESS
1st	time	1	.003 in.
2nd	time	2	.006 in.
3rd	time	4	.012 in.
4th	time	8	.024 in.
5th	time	16	.048 in.
6th	time	32	.096 in.
7th	time	64	.192 in.
8th	time	128	.384 in.
9th	time	256	.768 in.
10th	time	512	1.536 in.
11th	time	1024	3.072 in.
12th	time	2048	6.144 in.
13th	time	4096	12.288 in.
14th	time	8192	24.576 in.
15th	time	16384	49.152 in.
16th	time	32768	98.304 in.
17th	time	65536	196.608 in.
18th	time	131072	393.216 in.
19th	time	262144	786.432 in.
20th	time	524288	1572.864 in.
21st	time	1048576	3145.728 in.
22nd	time	2097152	6291.456 in.
23rd	time	4194304	12582.91 in.
24th	time	8388608	25165.82 in.
25th	time	16777216	50331.65 in.
26th	time	33554432	100663.3 in.
27th	time	67108864	201326.6 in.
28th	time	134217728	402653.2 in.
29th	time	268435456	805306.4 in.
30th	time	536870912	1610612.7 in.
31st	time	1073741824	3221225.5 in.
32nd	time	2147483648	6442450.9 in.

In other words,
the final pile of napkins is
      6,442,451 in. thick.
To change this to feet,
we must divide it by 12, obtaining:
      536,871 feet.
Or perhaps you would like
the answer in miles!
In that case, divide now by 5280,
since, as you know,
there are 5280 feet in 1 mile.
Thus we get:
nearly 102 miles!
Remember that 1 mile
is about 20 city blocks.
Now imagine a pile of napkins
over 100 miles in height!

Are you surprised again?
Did you get your answer by a "hunch"?
Or did you try to do it experimentally
by actually piling the napkins up?
Or did you calculate it as we did?

Let us discuss these various methods
a bit:

As regards a hunch,
we wish to make two points very clear:

(1) Some of our hunches are RIGHT
and some of them are WRONG.
The only way to tell
which is which
is to FOLLOW the hunch and
check it up.

(2) Scientists and mathematicians
also have hunches—
some of their best ideas
have been hunches;
but these do not become
respectable Science and
Mathematics
until they are
checked and double-checked.

This is one very essential difference
between the behavior of T. C. and
that of a scientist.
T. C. is apt to think that
if he is good at hunches sometimes,
he may rely on them always.
But the fact is that
EACH INDIVIDUAL HUNCH MUST BE
CHECKED AND DOUBLE–CHECKED!

Now as regards the experimental method:
this is generally known as
a very "practical" method:

"If you actually DO a thing,
 you cannot fail to get
 the right answer."
Often this is true,
but you can easily see that
in this particular problem
it is scarcely practical to
pile up napkins 100 miles high!
You would surely
hit the ceiling
if you tried it!
In short,
do not be too sure of
what is "practical" until
you have examined
the problem in question.

Finally,
we have the method of calculation:
this method was, as we saw,
by far the best in this case.
So let us NOT say that
Mathematics is IMPRACTICAL
whereas
doing things with your hands is
PRACTICAL.
This is SOMETIMES true,
but NOT ALWAYS!
If you think the calculations
were tedious,

27

we must point out:

(1) At least they were
not as tedious as
piling up the napkins
would have been!

(2) There is a much shorter way
to calculate the answer—
BUT
for this you need to know
a little MORE Mathematics:
namely,
a chapter in Mathematics
known as Logarithms.
We shall not explain it here,
for it is already explained in
any book on Algebra.
You can look it up.
And,
with a little effort,
thus learn a method which is
useful on MANY occasions.

Please remember that
it takes a little effort
to drive a car,
or to swim,
or to do almost anything.
But, if the result is worth while,

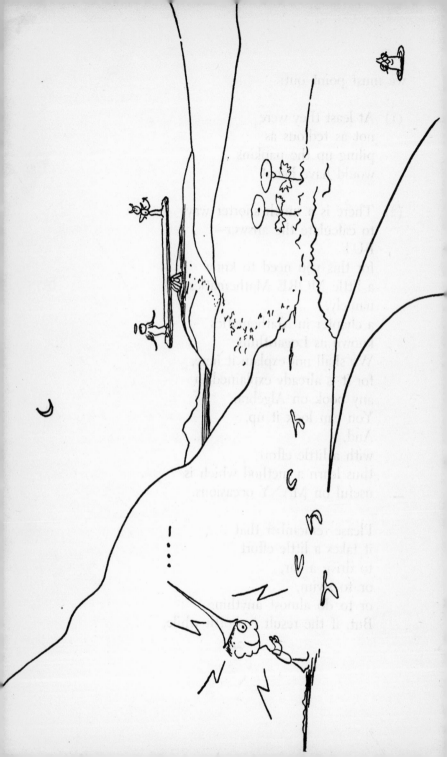

why growl at the effort?
After all,
the only way to make no effort at all
is to be dead!

The Moral: Wake up and LIVE!
             And
             follow your hunches and
             check them!

III. TISSUE–PAPER THINKING

Now that you are convinced that
we must be careful and
think more delicately,
you are ready to tackle
another question:

Suppose there were a steel band
fitting tightly around
the equator of the earth.
Now suppose that you remove it
and cut it at one place,
then splice in an additional piece
10 feet long,
so that the new band is
10 feet longer than the original one.
If you now replace it on the equator,
it would fit more loosely,
would it not?

The question is:

How large a space would there now be
between the band and the earth?
Would it be large enough for
(a) a man, 6 feet tall, to walk through,
(b) a man to crawl through
    on hands and knees,
(c) a piece of tissue paper
    to just slip through?

ANSWER BEFORE TURNING THE PAGE.

Did you say (c) is the right answer?
Perhaps this idea "flashed"
into your mind
because you felt that
10 feet could not make
much difference
in a band which was
thousands of miles long
in the first place.
Or perhaps you had learned
not to trust a "flash" too readily,
and decided to calculate
the answer
in the following way:
"Since the distance around
  the equator is 25,000 miles,
  dividing 25,000 miles into 10 feet
  gives a very small amount,
  and therefore
  I still think that
  (c) is the right answer."

But such a manipulation of numbers
can scarcely be called
"Calculating the answer."
For what justification is there
for dividing these numbers?
What is the THEORY behind this labor?
On more careful consideration
you must admit that

34

there is really no reason
for doing this.
In other words,
without a theory, a plan,
the mere mechanical manipulation
of the numbers in a problem
does not necessarily make sense
just because you are
using Arithmetic!

Now let us really examine
this problem
sensibly:
You probably know that
the circumference (C) of any circle
may be found by
multiplying its radius (R) by 2π,
where π is a Greek letter
(pronounced "pie")
and is a symbol whose
approximate value is 3 and 1/7.
Expressing this fact about a circle
more briefly,
we may say
$$C = 2\pi R.$$
And this is true of ANY circle,
no matter how large or how small.

Now if we increase the radius
by an amount x (see Fig. 1),

and make a new, larger, circle
whose radius is now $R + x$

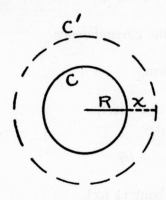

the new circumference would now be
$$C' = 2\pi (R + x)$$
would it not?
This may be written
$$C' = 2\pi R + 2\pi x.*$$
If we now compare this with
the value of C given above,
namely,
$$C = 2\pi R$$
we see that
C' is more than $2\pi R$
by an amount $2\pi x$.
In other words,

* Just as $5(2 + 7)$ may be written
 $5 \times 2 + 5 \times 7$, since in either case
 the answer is 45.

increasing the radius by x
increases the circumference by $2\pi x$
or by 6 and 2/7 times x.
Now, then,
if we increase the circumference
by 10 feet,
as in our problem,
we have

$$6\tfrac{2}{7}x = 10 \text{ feet,}$$

and consequently

$$x = 10 \div 6\tfrac{2}{7}$$

or

$$x = \text{about } 1\tfrac{1}{2} \text{ feet.}$$

That is to say,
an increase in the circumference
of 10 feet
results in an increase in the radius
of about 1 and 1/2 feet,
and consequently
(b) on page 33 is
the correct answer to our problem.
So you see,
you must not calculate mechanically,
like a robot.

Now that you have seen
a sensible method of solving
this type of problem,
try this one:
Suppose that you went
on a long walking tour

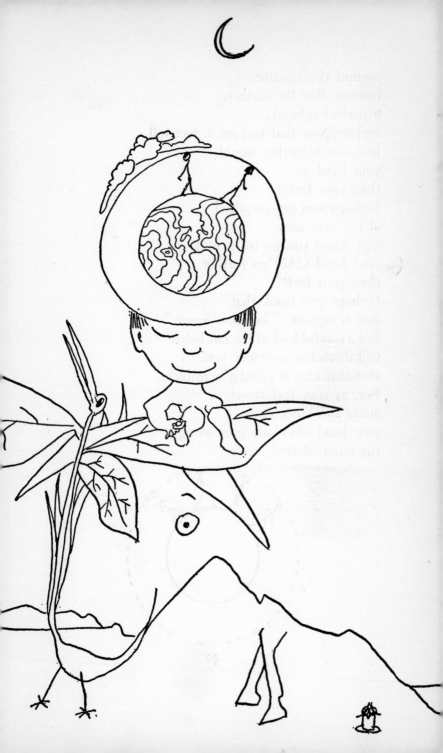

around the equator
(assume that the earth is
a perfect sphere),
and suppose that you are 6 feet tall,
how much further would
your head go
than your feet?!
Perhaps you are surprised
at the very idea
that when you go for a walk
your head CAN go further
than your feet!
Perhaps you think that
this is against "Common Sense."
But a careful look at the cut below
will doubtless convince you
that this idea is entirely sensible!
For, as your feet travel
along the inner circle,
your head obviously goes along
the outer, dotted, circle.

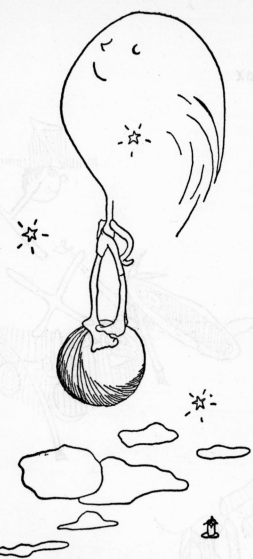

The Moral: Your head CAN go farther
than your feet!

IV. GENERALIZATION

No doubt you are aware of the fact
that the formula about the circle
in Chapter III is
"Algebra."
And perhaps you can guess from this
that one way in which
Algebra is different from Arithmetic
is:
Whereas in Arithmetic we do
one specific problem at a time,
in Algebra we give
a GENERAL rule for doing many
problems of a certain type.
Thus when we find
the area of a rectangle
whose base is 4 in. and altitude 2 in.
and get 8 square in.,
THIS IS ARITHMETIC.

But when we write
(1) $A = ab$
which says that
to find the area, A,
of ANY rectangle,
we must multiply the altitude, a,
by the base, b,
THIS IS ALGEBRA.

In other words,
Algebra is more GENERAL
than Arithmetic.
But perhaps you will say that
this is not much of a difference—
since in Arithmetic
we also have general rules,
but they are given in WORDS,
instead of in LETTERS as in (1).
Thus in Arithmetic we would say:
"To find the area of any rectangle
  multiply its altitude by its base,"
whereas in Algebra we say:
$$A = ab,$$
but, after all, you may feel that
this is merely a matter of
a convenient shorthand
rather than anything radically new.

Now the fact is that
it is not merely a question of

a convenient shorthand,
but
by writing formulas in this
very convenient symbolism—
especially when a formula is
much more complicated than
the one given above—
we are able to tell
AT A GLANCE
many interesting facts
which would be very difficult to
dig out from a complicated
statement in words.
And, furthermore,
when we learn to handle
the formulas,
we find that
we are able to solve problems
almost automatically
which would otherwise require
a great deal of hard thinking.
Just as,
when we learn to drive a car
we are able to "go places"
easily and pleasantly
instead of walking to them
with a great deal of effort.
And so you will see that
the more Mathematics we know
the EASIER life becomes,

for it is a TOOL with which
we can accomplish things
that we could not do at all
with our bare hands.
Thus Mathematics helps
our brains and hands and feet,
and can make
a race of supermen out of us.

Perhaps you will say:
"But I like to walk,
 I don't want to ride all the time.
 And I like to talk,
 I don't want to use
 abstract symbols all the time."
To which the answer is:
By all means enjoy yourself by
walking and talking,
but when you have a hard job to do,
be sure to avail yourself
of all possible tools,
for otherwise
you may find it impossible
to do it at all.

And so,
if you wish to be an engineer
and build bridges and things,
you must know Mathematics.
If you wish to figure out

46

how much money to put away now
so you may have
a comfortable income in your old age,
use a formula.
If you want to know
how much interest
you are REALLY paying
when you borrow money or
when you buy things on installments,
use a formula.
And, mind you, these are
algebraic formulas:
some of these problems
CANNOT be solved by using
Arithmetic alone!

You would be surprised to know
about some of the
remarkable and useful formulas
that would help you
if you would make
a little effort to
find out about them.

In fact
the trouble with the world today
is not that
we have too much Mathematics,
but that we do not yet have enough.
For, there are as yet

47

no powerful Mathematical methods in
Psychology,*
the Social Sciences,
and other important domains.
So that even the best workers
in those fields are,
figuratively speaking,
still using their bare hands,
and walking (rather than gliding) ,
and talking in ordinary language.
Perhaps, in these domains,
we ARE having fun all right,
but
we are getting nowhere
very fast,
for our wars are steadily becoming
bigger and better.

No doubt someone will say:
"But the war-makers
  DO use modern machinery which
  IS based on Mathematics.
  Science is really to blame for
  the success of Hitler,
  and therefore

* But see the work now being done,
  described in
  (1) "Mathematico-deductive Theory of
      Rote Learning" (Yale University Press).
  (2) L. L. Thurstone: "The Vectors of Mind"
      (University of Chicago Science Series).

48

it cannot possibly guide us to
the good life."
Now we hope to show here
that this is not so —
that Science and Mathematics can
not only protect us from
floods and lightning and disease
and other such physical dangers,
but have within them
a PHILOSOPHY which
can protect us from
the errors of our own
loose thinking.
And thus they can be
a veritable defense against
ALL evil—
a Totem Pole—
if we would but examine into them
carefully.

The Moral: Streamline your mind
with
Mathematics.

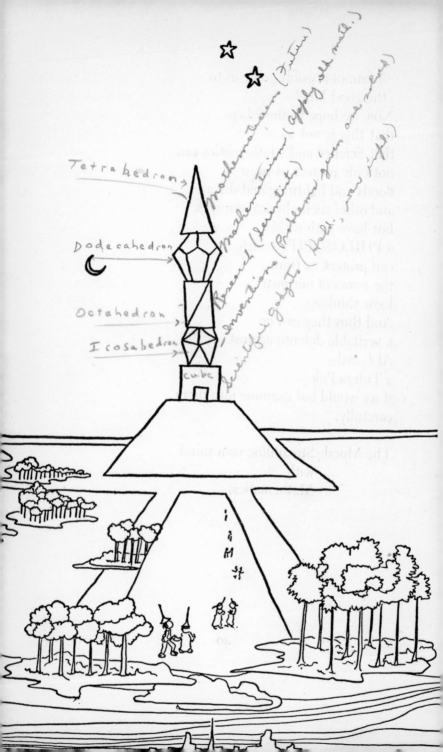

V. OUR TOTEM POLE

Let us symbolize our Totem Pole
by a column made up of
the five well-known regular solids,
as shown in the
drawing on the opposite page.
And let us think of the solids as
separate rooms, in each of which
a certain aspect of Science
is presented.
We shall now take you
on a guided tour of these rooms.

The First Floor, the CUBE,
contains all the scientific gadgets
with which we are all so familiar:
automobiles and refrigerators
and radios and airplanes,

and what seems like googols * of others.
This is of course also the room
in which you will find
tanks and bombers and
all the paraphernalia of war.
And this is why some people say
that Science is amoral,
since it produces,
with equal indifference,
the peaceful toys we enjoy so much
as well as the instruments of
destruction.
But these people have probably
never climbed the magic staircase

* A "googol" is a very large number,
namely, 10^{100}, or 1 with a
hundred zeros after it.
The term "googol" was
invented in fun by
a nephew of one of our
great American mathematicians,
Professor Edward Kasner of
Columbia University.
It is becoming a popular word
and will doubtless soon be
in the dictionary.
If you wish to know more about googol
and "googol plex,"
and many other interesting things,
see
"Mathematics and the Imagination"
by
Kasner and Newman (Simon and Schuster).

which leads from the gadget room
up into the other rooms of
our Totem Pole,
and are entirely unfamiliar with
their contents.

Let us therefore go up to
the Second Floor, the ICOSAHEDRON.
Here we find
a great industrial laboratory—
this is where the gadgets are
invented, tried out, manufactured;
the men working on this level
are not advertising and selling,
they are inventing.
They are told by
the people who hire them:
"We want a brighter light,
  a cheaper light,
  a more smoothly running car,
  an effective defroster for
  airplanes"—
and a thousand and one other things.
These research men do not
let their minds roam around
looking for interesting problems.
Their problems are handed to them,
and they must solve them within
a very reasonable length of time,
"or else."

53

Only "practical" men are
wanted here,
and not oversentimental ones.
For they may be told at any moment
to find effective ways of
killing people—
they must make
the best long-range guns,
the best poison gases,
the bombs which can
destroy the most people.
But you thought we promised
to get away from all this
as we climbed upward:
and yet this floor seems to have
even more diabolical possibilities
than the first one.
Perhaps if this second floor
were destroyed,
the war paraphernalia of
the first floor might
become obsolete and die out
of its own accord.
Are not these scientific inventors
the real devils after all?

But let us climb another flight
and see what goes on in
the OCTAHEDRON.
These men are doing research in

"Pure" Science.
They are not employed by
manufacturers or governments;
they are usually
professors at universities who
select their own problems because
they are interested in them.
They are not concerned with
any practical applications of
their ideas.
They are the theoretical men—
they ask the most "useless" questions.
For instance:
"What happens when you mix
  sugar and water and lemon?"
They call it "Sugar Hydrolysis"
instead of "Lemonade."
They study it in different solutions,
carefully varying the
relative amounts of
the substances involved,
and examine them with a polariscope
for days and days, and years and years,
keeping careful records
and publishing the results in
scientific journals.

Will these investigations
make them rich?
Or fat?

Or benefit them in any "practical" way?
Not at all.
Then why do they do it?
The answer is that
they are just driven by
Curiosity.
Once in a while they are consulted by
the men on the second floor,
but not so very often.
Usually they just
write up their results and
die without knowing whether
these will ever have any
practical use.
But the fact is that
their results ARE very often,
in the long run,
used by some second-floor scientist.
Indeed,
these second-story men find that
they must study the work of
the "pure" scientists
constantly.
But usually it is the work of
the "pure" ones of the past—
work which has already found its way
into the textbooks which
they have studied at the
institutes of technology,
rather than the current work

published in the journals of
Pure Science.

In fact,
the gentlemen working at
any given time on the
second and third floors
seem to have very little in common:
the second-floor men consider
the upper-floor men to be
"wild-eyed, absent-minded
  college professors.
  Some of them are perhaps just
  crackpots,
  who knows?
  It is safer to go to the theory of
  the established past,
  which has been duly
  tried and tested."
And, on the other hand,
the inhabitants of the third floor
look down upon the second-story men,
considering them to be mere
"hirelings and ignoramuses,"
and prefer to leave their results to
the second-story men of the FUTURE,
"who will be in a better position to
  appreciate them."
But, granting even that
this will be so,

what guarantee have we that
the uses that their ideas will find
WILL be decent, moral uses?
How do we know that
they are not storing up just
a lot of additional trouble for the
unfortunate future generations?
No,
let us climb up further,
and look at the Fourth Floor,
the DODECAHEDRON.

Here we find
the Mathematicians—
not the "Pure" Mathematicians,
for they live on the Fifth Floor,
in the TETRAHEDRON garret,
with the Modern Artists.
The fourth-floor mathematicians are
the ones who know the
Classical Mathematics of the past
and apply it to
the scientific findings of the
"Pure" Scientists of the third floor.
They take the scientific data
and organize it
and study it with
all the mathematical machinery
at their command.
If a second-story man ever

happens in on the fourth floor,
which is very rare,
he can hardly control his laughter.
These men seem to him to be
even more wild-eyed than those
on the floor below,
but the guide tells him:
"You ain't seen nothin' yet,
  wait till you see the Top Floor,
  the TETRAHEDRON."
At least on this fourth floor
you hear them mention
Geometry and Algebra and Calculus,
subjects you have heard about
in high school or in college.
But on that top floor,
they draw geometric figures on
doughnuts and pretzels
(no fooling!)
and on rubber sheets.
And they have up there
Algebras and Arithmetics in which
twice two is NOT four!
In which $3 + 2$ does NOT give
the same answer as $2 + 3$,
nor is 5×6 equal to 6×5!!
They are indeed fit companions for
the Modern Artists who
share the garret with them!
They are lucky if they can even

59

get a job!
And yet the connoisseurs say that
their work is
tremendously important for
the future.
Indeed,
if you trace back some of the
most practical and useful gadgets,
you will find that
if it had not been for a series of
"wild-eyed," "impractical" men,
these gadgets could not exist today.

As you will see in the next chapter.

Take the radio for example,
with all its variety of
concerts and important
broadcasts of all kinds.
Trace it to the second floor
and you will find that
many men on that floor have been
improving reception by inventing
better tubes and aerials, etc.
But all this could not have happened
had it not been for
a man named Marconi,
a second-story man,
who sent the first
crude radio messages.
And even his work
would have been impossible
had it not been for
another man, named Hertz.
who worked on the third floor,

and who proved that the very idea
of sending a wireless message
was actually possible,
since he demonstrated the existence of
electromagnetic waves.
But where did he get the idea of
even looking for these waves?
Why, from a fourth-floor man,
of course,
a man named Clerk Maxwell,
who first conceived the idea of waves in
an "electromagnetic field" and
applied the Calculus to it, obtaining
a set of differential equations
from which he declared
the consequence followed that
there MUST be
electromagnetic waves.
And, as we have already said,
Hertz subsequently proved
that he was right.
And, obviously,
Maxwell could not have done his job
had not Newton invented the Calculus.

And so it goes.

Take any gadget you like
and trace it back,
and you will find that

invariably you will have to
go up into
all the five floors
before you can have its complete story.

"But," you will say,
"you have not proved
  your initial point
  at all,
  since the same is true of
  tanks and bombers also,
  and therefore
  Science IS indifferent to
  Good and Evil,
  and IS amoral after all."

You will soon admit, however,
if you read again
the story we have told you,
more attentively this time,
that Science is trying to
tell you something else,
if you will but listen.
For instance,
go back and you will see that
in the little story of the radio,
there are
Americans,
Italians,
Germans,

63

Englishmen.
If you take the airplane,
you will also find
Russians,
Frenchmen,
and others.
In short you will be very much
impressed by the fact that
SCIENCE IS INTERNATIONAL,
that it is trying to tell us that
Hitler's racial theories are
utterly false.
It is also trying to tell us—
if we would only listen—
that co-operation is essential
for accomplishing things,
that it is really absurd
for the first- and second-story men
to laugh at those who live upstairs,
or for the latter
to look down upon the others.
For they are all needed
to do the job.
Is not this DEMOCRACY?

Thus we see that
Science is NOT AMORAL,
but has a PHILOSOPHY
to offer us,
provided that we do not
merely identify Science with

first-floor gadgets,
and thus
cut off its HEAD
(the upper floors!)
and stop its
BLOOD STREAM
(the interrelationship
between ALL the floors)!

And as we tell you more about
those strange
algebras and geometries we mentioned,
you will see that
Mathematics has many important
messages for us—
that it is trying to tell us that
VARIOUS mathematical systems
are possible,
that they are all man-made
and controllable by man,
and that
if you apply this idea
to the social world,
you will realize that
it is up to you
to build a good world if you want one—
that man has a great deal more
freedom and creative ability than
he is sometimes aware of.
The idea of a fixed "human nature"
that has us by the throat

ΔX

is just a fiction,
for
the activities on the top floor
are trying to tell us that
HUMAN NATURE HAS
INFINITE POSSIBILITIES.

In short,
it is not the guns and tanks
which are the real evils—
for a gun may be a great "good"
under certain circumstances.
But rather
such false IDEAS as
"Nationalism,"
"Dictatorship,"
narrow views of
"Human Nature,"
etc.,
are the real DEVILS.
Thus
FALSE IDEAS ARE MORE DANGEROUS
THAN GUNS!!
Guns and tanks are mere tools,
they may be used for good or evil.
But they are only
first-floor gadgets,
whereas the philosophy of science,
which comes from
a contemplation of all the floors,
and of their relationship

to each other,
has for us
unmistakable messages:
we must rise up above
the first and second floors
and realize that
these alone are
NOT SUFFICIENT
for the human race,
whose nature is so
beautifully revealed by
a study of Science as a whole,
which, as we have seen, has
Internationalism and Democracy
at its very heart.

We therefore advocate:
(1) A broader view of Science which
    enables us to appreciate the
    philosophy in it—
    Science as a whole,
    as seen in the Totem Pole,
    can really protect us from evil.
(2) A more appreciative attitude
    toward the top-floor men.
    By knowing how much we owe
    to the top-floor men of the past,
    we should stop treating them with
    the brutality with which

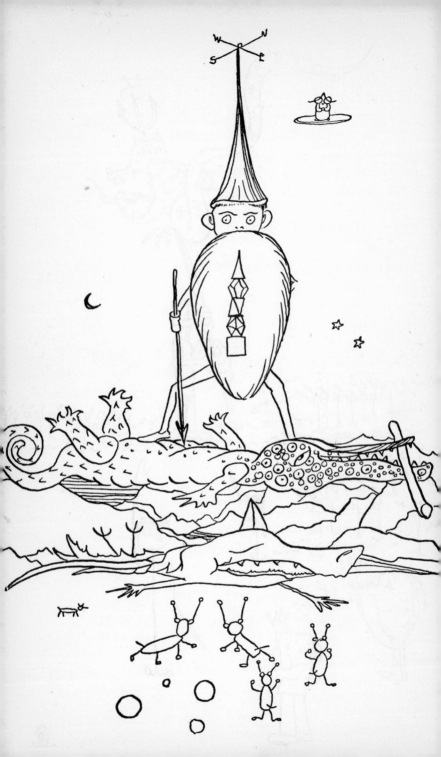

they have been treated
in the past—
just because they do not use
their energy to make themselves
physically comfortable.
And we should stop heckling them
by asking them:
"What is the practical use of
  what you are doing?"
or
"What does this mean for
  The Average Man?"
Since the truth is that
they themselves do not know.

Their work is as much
a "Natural Phenomenon" as
a natural oil well or
natural gas or
mountains or
rivers.
Let us give them the
freedom they need
to do what is in them to do.
Let us turn their garret into a
penthouse,
and marvel at their
strange products.
Perhaps some day we shall find
a "practical" use for them,

as has so often happened
in the past.
And besides,
the philosophical implications
of their work
already make them
invaluable to us NOW—
as we shall see.

The Moral: Oh, listen to the
Totem Pole!

VII. ABSTRACTION

You saw in Chapter IV that
GENERALIZATION is
one of the principal advantages
that Algebra has over Arithmetic.
In fact GENERALIZATION is
one of the fundamental methods
of obtaining new results
in all of Mathematics.

Perhaps someone will say:
"But generalization is not the
   private property of mathematicians,
   every man knows that
   all women are silly.
   Every woman knows that
   all men are fools.
   And everyone knows that
   all Jews are
   bankers AND Communists."
We need hardly say
that these generalizations are
NOT VALID,

whereas in Mathematics
we take pride in making
our generalizations with
MUCH GREATER CARE.

Now, in Geometry, as you know,
we deal with
the relationships between
points, lines, planes, and so on,
and study the properties of
various figures
(triangles, circles, etc.)
and
various solids
(prisms, spheres, etc.) .
And, as you also know,
we draw diagrams of plane figures
on a blackboard or a piece of paper,
and make models of
three-dimensional objects,
to help us visualize
the things we are discussing.
But of course you realize that
a point drawn on a blackboard
with chalk,
or on a piece of paper
with even the finest pencil or pen,
is much too large for
a mathematical point,
which is supposed to have

75

no dimensions at all—
no length, no breadth, no thickness.
And, similarly,
a circle drawn with
even the best instruments
is only a crude representation
of a mathematical circle.
Thus
the things with which we deal in
Geometry
are ABSTRACTIONS of actual things
in the physical world.
And just because they ARE
abstractions,
they are therefore
EXACT instead of APPROXIMATE.
For example,
every point on the circumference of
a mathematical circle
is at EXACTLY the same distance
from the center.
But you might say:
"Even if they are exact,
    what good are they when
    they exist only in the mind?"
You will soon see
what a mathematician can do
with abstractions and
how they can be applied
to the actual world.

In fact,
this power to ABSTRACT is
one of the outstanding characteristics
of human beings as
compared with other animals.
And this power is used not only
by mathematicians,
but also by
artists, musicians, poets,
and all other "human" beings.
Perhaps some day
we shall measure
a person's "human-ness" by
his power to abstract
rather than by the I.Q.
For a person who can
be loyal to such
abstract concepts as
truth, justice, freedom, reason,
rather than to
an individual or a place,
has the loyalty of a human being
rather than that of a dog.
Please do not think that
we are using the word "dog"
in a disparaging sense,
for they are very dear animals.
(Remember that you must not be
a Conclusion-Jumper!)
But still they are animals and

not human beings.

But what are
"Truth,"
"Justice,"
"Freedom,"
"Reason,"
etc.?
Do these words really mean anything?
And how can we be loyal to them
if their meaning is not clear?
Are they not just "fakes,"
invented so that
some people can make slaves of others
by fooling them with such
meaningless abstractions?
Now you will see,
when you have finished this
little book,
that these concepts
"Truth," "Freedom," "Reason," etc.,
will become much clearer when
we examine into what is meant by
"Mathematical Truth,"
what kind of "Freedom" we have
in Mathematics,
what is considered good "Reason"
in Mathematics,
and so on.

You will see that
as mathematicians have been
gradually forced to consider
the fundamentals of Mathematics,
they have been obliged
to consider the very nature
of human thinking—
both its powers and
its limitations.
For instance,
what is the nature of a "proof"
by human beings
for human beings?

And, of course,
this has a definite bearing on:
"What are we humans anyway?
  What is the best that
  we can expect of ourselves?"

The Moral: Be a man—not a mouse.

VIII. "DEFINE YOUR TERMS"

So far then
we have said that
GENERALIZATION and ABSTRACTION
are very fundamental and useful
human concepts.
And we must emphasize the fact that
Mathematics is not the only domain
in which these concepts are used.
For example,
a great symphony
does not have specific words
like a popular song,
and thus it abstracts an emotion
rather than giving
a particular instance of
an emotion,
and therefore has
a wider application.

Similarly,
a great portrait
is more abstract than a photograph

81

because it does not represent
the person as he looks at a
particular moment,
but abstracts what the artist
considers to be
the essential character of
his subject.
Perhaps someone will say:
"I agree that this kind of
 abstraction
 is good,
 for I admit that a great portrait
 has a wider scope
 than a photograph.
 But what about these MODERNS
 who abstract to a degree
 where the subject is no longer
 recognizable at all?"
But let us not discuss
the MODERNS here,
for remember that
this Part I is called
"The Old," not "The New."
We shall discuss the MODERNS
in Part II.
For the present
we merely wish to point out
that
the concepts of
GENERALIZATION and ABSTRACTION

in Mathematics,
as well as in Art, etc.,
have an "OLD" and a "NEW" aspect.
And the uses of them described above
belong to the "OLD" in Mathematics
just as portrait painting is
an "OLD" form of abstraction
in painting.
And the "NEW" in Mathematics,
as well as in Art,
may also sound bizarre
to the uninitiated.
For example,
as we have said before,
to a modern mathematician
2×2 does not have to be 4!!
But do not let this frighten you,
for when you have read Part II,
you will have become
so broad-minded
(we hope)
that such modern ideas
will seem just as reasonable
as anything you believe today.

But let us not anticipate;
and continue with our story.

What other fundamental ideas
do we find in Mathematics?

No doubt many of you will say:
"Surely you will discuss the fact
  that Mathematics is a domain
  in which
  we prove everything,
  in which we carefully
  define all our terms,
  so that we know what
  we are talking about.
  And the moral of this
  will doubtless be that
  we should learn
  to define all our terms
  in ANY argument,
  and thus use
  the mathematical method
  as a model."

Well,
we are sorry to disappoint you,
but we must tell you that
even Euclid,
as far back as 300 B.C.,
already realized that
it is IMPOSSIBLE to
define all of our terms or
to prove everything,
even in Mathematics!
For, you see,
since in a proof

84

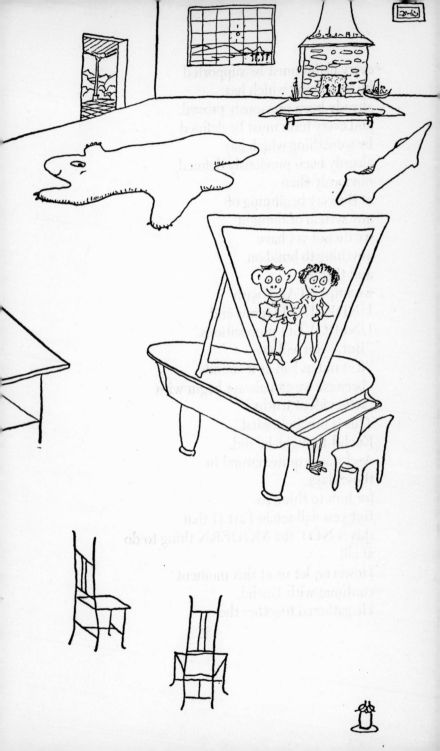

every claim must be supported
by something which has
already been previously proved,
and every term must be defined
by something which has
already been previously defined,
obviously then
at the very beginning of
any system of thought
we do not yet have
anything to build on
and therefore
we must START with
UNDEFINED terms and
UNPROVED propositions.
"But," you will say,
"it is not as bad as it sounds,
  because we can always begin with
  self-evident truths."
This is precisely what
Euclid thought he did.
And it was quite natural in
those days
for him to think so.
But you will see in Part II that
this is NOT the MODERN thing to do
at all!
However, let us at this moment
continue with Euclid.
He gathered together the

86

geometric knowledge of his time,
and arranged it
not just in a hodge-podge manner,
but, as we said above,
he started with what he thought were
self-evident truths
and then proceeded to
PROVE all the rest by
LOGIC.
A splendid idea, as you will admit.
And his system has served
as a model
ever since.
But, as we promised above,
you will see in Part II
the very fundamental changes
which mathematicians have been
obliged to make in
Euclid's system.
And, therefore,
with all due respect to Euclid,
we must not slavishly follow him
TODAY,
as so many of our Geometry texts
do!

The Moral: Progress is made by
respecting tradition
without slavishly
following it
100 per cent!

IX. A WEDDING

If you look back over the history
of the human race,*
you will find that
many useful things from
Arithmetic and Algebra
were known as far back as 4000 B.C.;
that Geometry reached
a high stage of development in
the work of Euclid, about 300 B.C.
Since then
many more things have happened
in Mathematics:

(1) Algebra and Geometry have both
     been developed further.
(2) They have been COMBINED into
     a new branch of Mathematics
     known as Analytic Geometry—
     by Descartes in the 17th century.
(3) Many new Algebras and

* In this connection read:
  "The Development of Mathematics" by
  E. T. Bell (McGraw-Hill).

90

many new Geometries
have been developed.
(4) The FUNDAMENTAL IDEAS of
all Mathematics
have been carefully examined.
(5) Logic has been inspected
and new logics have arisen.
(6) New applications of Mathematics
to the study of the universe
have been made.
(7) And, as a result of all this,
mathematicians have become
much wiser,
much more sophisticated.
And their "common sense" has
become so enlightened that
they cannot help but look upon
the more usual common sense
of T.C. Mits as
an adult looks upon the
common sense of a young child
who thinks that
every man is his daddy.

Of course it takes a great deal of
powerful thinking to become
a great mathematician,
but we believe it is possible
to give T.C. a glimpse
into the results,

without asking him to become
a mathematician himself.
In this chapter we want to tell him
about the wonderful
17th century wedding
mentioned in (2) above—
and about the offspring.
Descartes conceived the idea of
associating Algebra and Geometry
in the following manner:
If we draw two perpendicular lines,
X and Y,
as shown in the next figure,
thus dividing the plane of the paper
into four "quadrants," I, II, III, IV,
we can associate
every point in the plane with
a pair of numbers, thus:

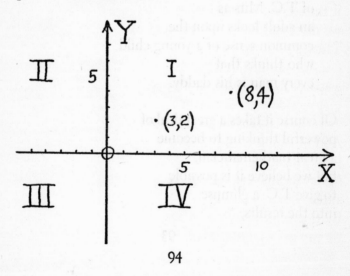

(3,2) designates the point which
is located 3 spaces to the right of O
and two spaces up.
Similarly (8,4) is the point
located 8 spaces to the right
and 4 up.
Note that the first of
the two numbers gives
the distance along the X axis,
and the second number of the pair
gives the distance parallel to
the Y axis.
And
if the first number is negative,
like −2,
we must go to the LEFT on the X axis
instead of to the right.
And, similarly,
if the second number is negative
we must go DOWN instead of up.
Thus
(−4,−5) designates a point
4 spaces to the left of O and
5 spaces down
(see the diagram on page 96).
And of course
(−4,5) means 4 to the left and 5 up,
(4,−5) means 4 to the right and 5 down,
and so on.
By this simple device
we can get a picture which

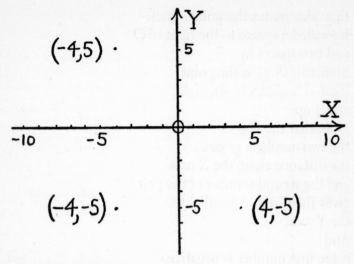

(-4,5) ·

(-4,-5) ·   · (4,-5)

gives us the information that is
often contained in
columns of numbers,
and gives us this information
much more vividly.
For example,
if the temperatures at a given place
during a certain day are:

Time	Temperature
2 A.M.	$-4°$
5 A.M.	$0°$
6 A.M.	$3°$
9 A.M.	$5°$
11 A.M.	$8°$
6 P.M.	$1°$
9 P.M.	$-3°$

96

they may be represented
"graphically" thus:

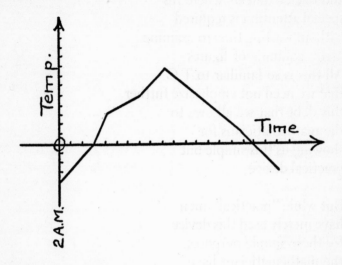

Graphs of this type are
doubtless familiar to you,
for the "practical businessman"
has seen the tremendous advantage
that this pictorial kind of
representation has for
his business,
in advertising,
in examining his volume of business,
and so on and so on.
A physician making his daily visit
to a hospital
can walk through a ward and see

97

each patient's temperature chart
at a glance,
and decide quickly where his
special attention is required
without wasting time to examine
many columns of figures.
All this is so familiar to T.C.
that we need not emphasize further
this debt that we all owe to
the mathematicians for
showing us this simple but
practical device.

But while "practical" men
have merely used this device
for these simple purposes,
the mathematicians have,
by playing with the device itself,
put it to infinitely greater use.
We shall not give you here
the details of how
the mathematicians developed
this simple device of a "graph"
into a branch of Mathematics
known as Analytic Geometry,
without which
we would not have had
Newton's Calculus with its
tremendously important applications
to

Engineering,
Physics,
Chemistry,
with the resulting benefits to us
in
Transportation via
railroads and ships and planes;
in
Communication via
telephone and telegraph and radio,
and all the other benefits in
diet,
health,
air-conditioning,
etc., etc.
For these are all described in
other books,
and there is no need to
repeat these stories here.

The Moral: Why not read some of
these stories in
your spare time?

X. THE OFFSPRING

We just want to indicate briefly here
one major idea of Newton's Calculus:

Suppose you are taking a trip
in an automobile
and traveling at a steady rate of
40 miles an hour.
How far can you go in 2 hours?
Obviously the simple formula
(2) $d = rt$
(distance = rate × time)
will give you a quick answer.
But suppose that
your rate is not constant;
you can easily see that
this formula will no longer work.
And since we often have need
for formulas which will apply
to motions in which
the rate is not constant,

let us see how this can be done.

To do this easily,
let us first
plot the graph of equation (2)
for the case when $r = 40$,
namely
(3) $d = 40t$.
We first make a table,
by giving t any values we please,
and calculating from (3)
the corresponding values of d:

t	d		
0	0	3	120
1	40	4	160
2	80	5	200

and then plot these points:

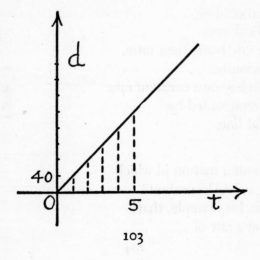

Now since equation (2)
may be written

$$r = \frac{d}{t}$$

(to find the rate,
divide distance by time),
we see from the graph that
the rate may be found by
dividing the value of
any dotted line
(which represents distance traveled)
by the corresponding value of t.

And so the graph on page 103
completely shows
the motion in question,
the time being shown
along the horizontal axis,
the distance along
the vertical axis,
and the rate being their ratio.
And obviously,
a motion having a constant rate
will be represented by
a straight line.

Now,
what about a motion in which
the rate is NOT constant?
Suppose, for example, that
you go at a rate of

104

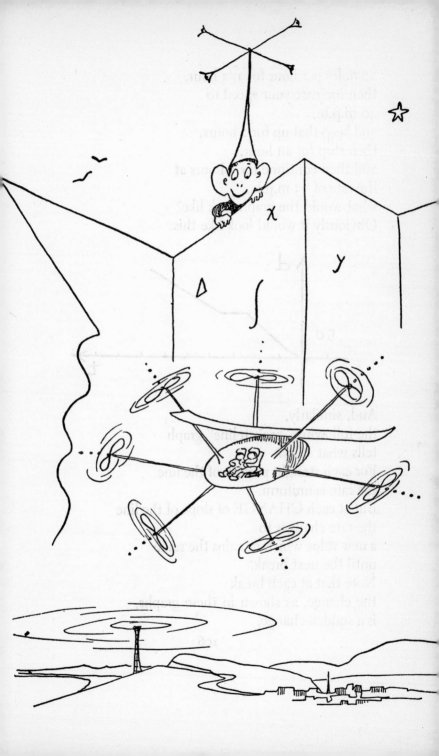

20 miles per hour for 1/2 hour,
then increase your speed to
40 m.p.h.,
and keep that up for 2 hours,
then stop for an hour,
and then continue for 3 hours at
the rate of 35 m.p.h.,
what would the graph look like?
Obviously it would look like this:

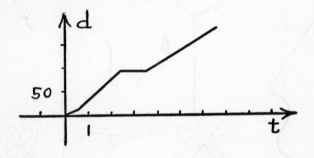

And, similarly,
the following "broken line" graph
tells what story?
For each straight portion of the line
the rate is uniform.
But at each CHANGE of slope of the line
the rate changes to
a new value which remains the same
until the next break.
Note that at each break
the change, as shown in these graphs,
is a sudden change,

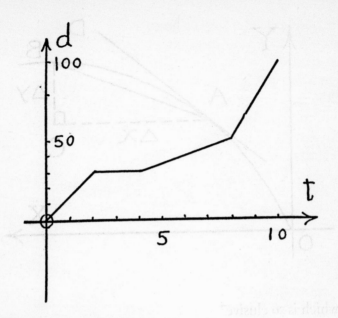

no allowance being shown for the
process of accelerating or
slowing down.

To show this process,
we must have a CURVE as shown
on page 108,
where x is the time and
y the distance covered.
Here any particular rate is
not kept up for an appreciable time
but is CHANGING ALL the time.

How can we now "catch" a thing

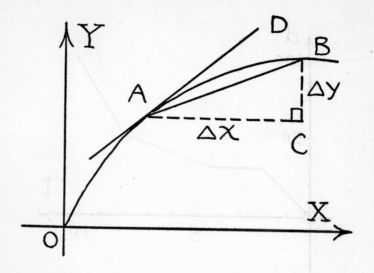

which is so elusive?
That was the problem solved by
the Calculus:
Suppose first that
the motion from A to B were
a uniform motion instead of
an accelerated one.
Then it would be represented by
the straight line AB instead of
the curve AB.
And it would show that
in time AC
the distance BC was covered,
at a constant rate equal to
$$BC/AC.$$

Now as you take the point B
nearer and nearer to A,
the straight line AB approaches
more and more to the line AD
which is tangent to the curve
at point A.
Thus we may say that
the actual rate at A is
the "limit" of BC/AC.
And whereas
this rate lasts only an instant,
(for as soon as you get away from A
the slope of the tangent line is
obviously different),
still
we can "catch" it and
express it mathematically
(and thus be able to work with it).
Thus,
if we represent AC by Δx
(read "delta x"),
which simply means the
difference in the x-value
from A to B,
and BC by Δy,
then,
as B approaches A,
this ratio $\Delta y/\Delta x$ approaches
a limiting value.
This limiting value of $\Delta y/\Delta x$ is

represented by dy/dx.
And so we have
$$dy/dx = r,$$
the rate AT THE POINT A —
and r of course changes
from point to point.

Now,
if we know
the equation of the original curve,
the Calculus gives us
the necessary machinery
(called "Differentiation")
by which we may find
$$dy/dx$$
at any point.
And, vice versa,
if we know the value of dy/dx,
that is, if we know the
"Differential Equation,"
we can find,
by means of the Calculus
(by "Integration") ,
the equation of the original curve.

Now in most physical problems,
in this ever-changing world,
the idea is
to set up a differential equation
which represents what is happening

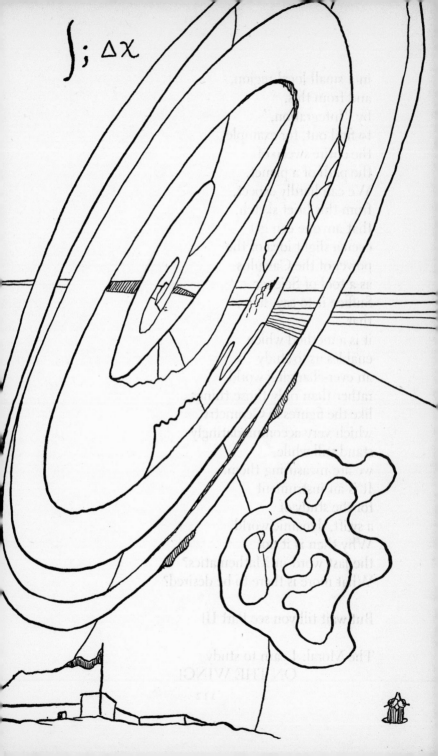

in a small local region,
and from this,
by "Integration,"
to find out, for example,
the entire sweep of
the path of a planet.
We can hardly expect,
from this brief sketch,
that anyone can get
even a slight idea of the
power of the Calculus
as a tool of Science.
Suffice it to say, here,
that
it is a method which
enables us to study
an ever-changing world,
rather than only those things,
like the figures in Geometry,
which very accommodatingly
stand still while
we are measuring them.
It is an instrument
for the study of
a swift, dynamic world.
Why then is it not
the last word in Mathematics?
What more is there to be desired?

But wait till you see Part II!

The Moral: Learn to study
　　　　　ON THE WING!

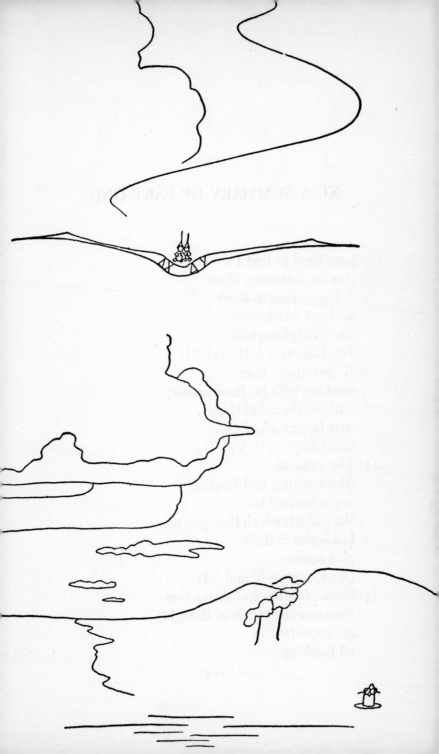

XI. A SUMMARY OF PART ONE

We have tried in Part I to
give you the following ideas:
  (1) A man trying to think
     without Mathematics is
     like a helpless child
     (see Chapters I, II, and III).
  (2) A "practical" man
     working with his hands alone,
     without the aid of theory,
     may be just a fool
     (see Chapters II, V, VI).
  (3) The value of
     Mathematics and Science
     is not limited to
     the gadgets which they give us,
     but is also in their
     philosophy
     (see Chapters V and VI).
  (4) Generalization and abstraction
     (two powerful tools of thought)
     are important in
     all thinking.

You cannot really think
without them.
But you must learn to use them
properly.
If used carelessly
they are "dynamite" and
may blow you up!
(see page 75 and page 67).

(5) Do not always demand a
"Yes" or "No" answer.
For example:
"Shall we cling to the traditions
of our great forefathers?
Yes or no?"
The history of Mathematics shows
just how much of Euclid
we must keep
and how much we must discard.
You will see this in some detail
in Part II.
But outside of Mathematics,
in the social studies,
you will hear people quoting
blindly:
quoting the Constitution,
quoting Karl Marx,
quoting Theodore Roosevelt,*

* By the way,
the men who wrote the Constitution,
as well as other men so often quoted,
would be horrified at some of the
applications made by their disciples.
BEWARE OF DISCIPLES!

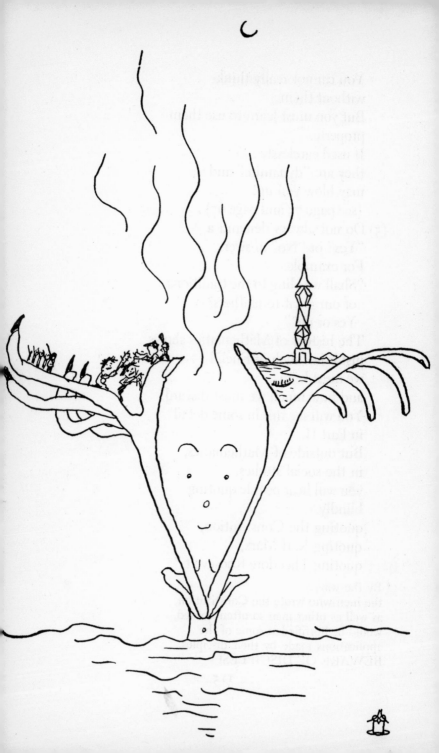

with the implication that
you must either
completely accept or
completely reject.
In Mathematics, however,
we do not just quote authority.
We say:
"In the light of our knowledge today,
  Euclid was right in this and
  wrong in that."
And this is a wholesome way
to look at the past.
It is partly good and partly bad;
we must select,
in the best light of
our knowledge now.
(6) Do not jump at conclusions
    (see Chapters I, II, and III) .
(7) Do not rule out hunches
    because they are sometimes wrong.
    Read some of the original
    writings of Faraday or
    other great scientists—
    you will be surprised to find
    how much of their work
    started as a "hunch."
BUT
  (8) Do not think all
    your hunches are wonderful!
    Some of them may be terrible!

Follow them up cautiously!
Encourage them but
watch them!

(9) Try to judge
statements and theories
in the light of
important long-time activities
of the human race—
like Science or
Mathematics or
Art.
They reveal "human nature"
better than anything else.
In them you will see that
Internationalism and Democracy
are very deep in the human spirit
(see Chapter VI).

(10) And so you see that
Mathematics is not for
the engineer only,
or only for someone who
needs its formulas.
It is a way of thinking,
a way of life,
VERY IMPORTANT FOR EVERYONE.

(11) Most courses in Mathematics
do not leave time
to consider all these things.
They are too full of technique.
We MUST stop now and then

from the manipulation of
techniques
to see what
general ideas we can get from them,
which will be useful for
ALL of us.

PART II

THE NEW

XII. A NEW EDUCATION

And so you know that
Algebra is a sort of
Generalized Arithmetic
by which more difficult problems
may be solved.
That Geometry is
not only the study of
various figures in
two and three dimensions,
but is also
a sample science,
the entire structure of which
is built up from
a few basic postulates—
and is therefore a "model"
for any system of thought.
That Analytic Geometry is
a combination of
Algebra and Geometry which
has proved extremely useful.

And that Calculus is
a powerful instrument for
the study of
our DYNAMIC world.

You know also that
Mathematics is useful not only
as a technique,
but also as a sample of
a method of thinking:
it is clear,
precise,
brief,
many-sided.
That a THOUGHTFUL study of
even a little Mathematics
can throw much light on
many controversies,
even with very little use of
mathematical technique
(see the summary of Part I) .

Perhaps you may say:
"What more can we ask?"

But the fact is that
all the branches of Mathematics
mentioned in Part I
had been discovered
by the time of Newton,

who lived from 1642 to 1727.
And it was he who
invented the Calculus.
Analytic Geometry dates from Descartes,
about 1637.
Euclid goes back to
about 300 B.C.
And a good deal of the Algebra
which is studied in
high school and college
is spread out
from as far back as
about 3000 B.C. to
the time of Newton.

Thus,
the knowledge of Mathematics
of the average college graduate
stops with what was known
about 300 years ago!
And yet
more Mathematics has been invented
in the last 100 years
than in all the previous centuries
taken together!
If the same were true about
the study of Physics,
the average college graduate would
never even have heard of
an airplane or

an automobile or
a radio,
etc., etc.
Such a situation in Physics
would never have been tolerated.

Why then is it tolerated in
Mathematics?
Perhaps MODERN Mathematics is
so difficult that
it can be understood only by
a few rare souls?
Not at all!
Of course it took
a few VERY RARE souls indeed
to CREATE it,
But these new results are
no harder to understand than
any of the older Mathematics.

Perhaps it is just inertia
on the part of some educators?
And T.C.,
not being aware of
what he is missing,
does not clamor for it!

We feel that he can get even more of
an intelligent, general outlook on
life

126

from the MODERN ideas than
from the older ones!

Read the next few chapters and
see if you agree with us.

XIII. COMMON SENSE

As we have already said,
one of the chief values of
the study of Geometry
lies in the fact that
it is a model for
any science
or for
any system of thought,
since it starts with
a few basic ideas
from which all the other ideas or
"propositions"
are derived by logic.

Now Euclid regarded these basic ideas
as "self-evident truths";
and
some of them seemed to him
so "self-evident" that
he did not think it necessary

even to mention them.
For example,
he thought it was so obvious
what is meant by
the "inside" and the "outside" of
a triangle
that he did not bother to
define them.
And no doubt T.C. is thinking
this minute
that it is only "common sense,"
and that any fool can SEE
which is which
at a glance.

But he will soon see the trouble
this "common sense" caused,
for we shall now show him that
it led to the following absurd
proposition:
"If a triangle is NOT isosceles,
  then it MUST be isosceles"!
(You remember of course that
an isosceles triangle is one which
has two of its sides equal.)
In order to follow the proof
you may have to recall
some of your high-school Geometry—
but that cannot hurt anyone, much.

Given, then, that
AB does NOT equal AC;
we shall now prove that
THEREFORE
AB DOES equal AC!

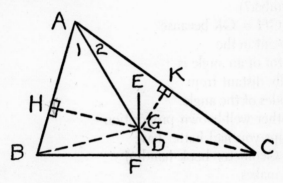

First draw AD so that angle 1 = angle 2;
and let FE be
the perpendicular bisector of BC.
Now, if the triangle were isosceles,
AD and EF would be
one and the same line,
but since the triangle is NOT
isosceles,
AD and EF must intersect.
Let us call their point of intersection G.
And now draw BG and CG;
also draw GH perpendicular to AB,
and GK perpendicular to AC.
Now BG = CG because
any point in the

perpendicular bisector of a line
is equally distant from
the ends of the line.
(This is a well-known proposition
in Geometry,
remember?)
Also $GH = GK$ because
any point in the
bisector of an angle is
equally distant from
the sides of the angle.
(Another well-known proposition—
perhaps you had better have
your Geometry book handy!)
This makes
triangle BGH congruent to
triangle CGK,
since
two right triangles are congruent if
the hypotenuse and a leg of one
are equal respectively to
the hypotenuse and a leg of the other.
(Another call for that Geometry book!)
Therefore $BH = CK$ (1)
for
corresponding parts of
congruent figures
are equal.
Similarly
triangle AGH is congruent to
triangle AGK,

and consequently, $AH = AK$. (2)
Hence, adding (1) and (2) above,
we get $AB = AC$,
showing that
two UNEQUAL sides of a triangle
MUST be EQUAL!

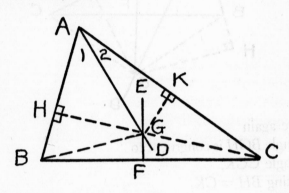

You probably do not like this result
any more than
the mathematicians did.
Now,
if you remember your Geometry
pretty well,
you will immediately say that
you know just where the trouble is:
namely,
that AD and EF intersect all right
BUT
NOT as shown in the diagram—
that they really intersect

133

outside the triangle,
like this:

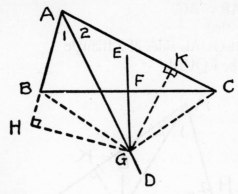

Here again
triangle *BGH* is congruent to
triangle *CGK*,
making *BH* = *CK*.
And
triangle *AGH* is congruent to
triangle *AGK*
so that *AH* = *AK*.
But now this does NOT make *AB* = *AC*,
since *AH* + *BH* NOW does NOT equal *AB*,
as it did before
(although *AK* + *KC* still equals *AC*
as before).
Hence now
we do NOT get the absurd conclusion
we got before.

BUT
we cannot let you off so easily!

Because
you are using the DIAGRAM
to prove your point,
instead of LOGIC!
Perhaps you will demand to know
"What is the difference?!"

Well,
the difference is that
diagrams are NEVER used as
evidence in Geometry.
Why?
Because Geometry is not
that kind of subject.
It is a subject in which
the theorems are derived from
the basic postulates
by means of LOGIC.
And if there is no definition given
of "outside" and "inside" of
a triangle,
no argument can be based on
such a nonexistent definition.
Do you think this is just
quibbling?
But mathematicians have been fooled
by this sort of thing before,
and are more cautious now—
and that is what we recommend
to T.C. also.
For how would he like to

be accused of a crime which
is not even recorded in
a law book?
Would he not then appreciate
a lawyer who would argue
that there is no validity in
a "tacit" law?

Thus mathematicians too
have learned by hard experience
not to base an argument on
"tacit" assumptions.

Perhaps this brings to your mind
cases from modern psychology,
in which much damage is done
to an individual's nervous system
by "subconscious" ideas,
which, if brought up
into consciousness,
can be treated and
eliminated as a source of difficulty.

Thus, one modern trend seems to be
to turn the light on
our subconscious thoughts and
rid ourselves of the
prejudices and false thinking
which may be due to them.

The Moral: Be REASONABLE
            by bringing to light
            your "tacit" ideas.

XIV. FREEDOM AND LICENSE

As we just saw,
one of the tasks of
modern mathematical research
has been
to go back to Euclid,
bring to light his
"tacit" assumptions,
and make the kind of "phoney proof"
shown in the last chapter
IMPOSSIBLE,
at least in Euclidean Geometry.
And is it not up to us
to take a leaf from
the mathematician's book,
and profit by his experience,
by trying to make
"phoney proof" impossible also in
other domains of argumentation,
by not permitting any
"tacit" assumptions?

But this is not all!

What about the
"self-evident truths"
which were not tacit,
but were expressly stated?

Well,
one of these
even at that time
did not seem so "self-evident":
namely, the one which says that
"Through a given point,
  which is not on a given line,
  one and only one line can be drawn
  parallel to the given line" —
known as the famous
"parallel postulate."
Not considering it to be
"self-evident,"
Euclid tried to prove it,
but without success,
and finally listed it as
"self-evident,"
although he did not feel
so good about it.

Then,
for several hundred years,
outstanding mathematicians again

tried to prove it,
but again without success.
FINALLY,
but NOT UNTIL 1826,
a remarkable thing happened.
The idea dawned in the minds of
several mathematicians at the
same time
(Lobachevsky, Bolyai, Gauss)
that
not only is this statement
NOT "SELF–EVIDENT,"
but in what sense is it
"true" at all?
And so they undertook
to see what would happen if
they assumed that
"Through a given point
  (not on a given line),
  TWO lines could be drawn
  parallel to the given line,
  one to the right and
  a different one to the left."

Perhaps T.C. will immediately object
and say,
"But this is impossible";
and will get all excited and
draw a diagram like this:

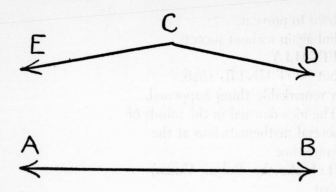

and say:
"Can't you see that
if you draw
two distinct lines through C,
as shown,
neither one of them can be
parallel to *AB*?
For *CD* will meet *AB* somewhere
to the right,
and *CE* will meet it on the left.
A little common sense
shows this so clearly."

But we must warn T.C. against
that "common sense" of his;
not that it isn't a good thing,
but he must use it with
SO MUCH MORE CAUTION
than he realizes

142

(don't forget Chapters I, II, III
in Part I,
and page 93) .
And we must also warn him again
against this reckless use of diagrams,
and compel him to
stick to LOGIC
as his SAFEST weapon for
clear thinking.

Now when these three
very intelligent men,
mentioned above,
looked into the matter
(quite independently of each other,
by the way),
they found that
no logical fallacy resulted from
their strange assumption,
but that they got
an entirely DIFFERENT Geometry!
A queer-sounding Geometry in which
the sum of the three angles of
a triangle
was no longer 180°,
in which
the famous Pythagorean Theorem
was no longer true,
and yet
the LOGIC was PERFECT!

143

Well, you may say,
"So what?
  A couple of wild-eyed,
  impractical mathematicians
  go haywire,
  lose all common sense,
  fool around with their
  silly logic,
  get ridiculous results,
  and I, T. C.,
  should get excited about it?"
Well, T.C.,
what would you say
if we should tell you that
in 1868
a man named Beltrami found that
all this stuff was not just
fantastic nonsense
but actually applied on a surface
called a "pseudo-sphere"?!
And he then understood that
whereas
good old Euclidean Geometry
applies on a flat surface,
like an ordinary blackboard or
a piece of paper,
that other Geometries are needed
for other surfaces,
and that it is all
very sensible.

Just as, for example,
the Geometry on the
surface of the earth
is also non-Euclidean,*
since here too
the sum of the angles of a triangle
does NOT equal 180°:
thus consider the triangle
formed by an arc of the equator and
the parts of two meridians drawn
from the north pole to
the ends of this arc;
the two base angles here are
each equal to 90°,
so that all three angles together
do NOT add up to 180°.

What, then, about
those "self-evident truths"?
Apparently BOTH
Euclid's parallel postulate AND
the non-Euclidean parallel postulate
mentioned above
(permitting TWO parallels through
a given point)
are equally true,

* See "Non-Euclidean Geometry"
  by Hugh Gray and Lillian R. Lieber
  (Galois Institute Press).

145

one of them applying on one surface,
the other on a different surface!

And so gradually
the mathematicians have been forced
to the position that
in Mathematics,
instead of regarding
the fundamental postulates as
"self-evident truths"—
as Euclid did—
it makes more sense,
in the light of the experience
described above,
to regard them as
mere ASSUMPTIONS.

In other words,
the mathematician now would say
if I assume some things
and then derive from them
other things by the use of
Logic,
WITHOUT CONTRADICTING myself,
that is all I ask.
For, fundamentally,
I am not even concerned with
the question of actually
finding a surface for which
a certain Geometry holds;

146

for my job is to find out
what straight thinking can do.
In a way it is really a
psychological problem.
And I find that
in some ways
I have a good deal of freedom;
and, in other ways,
I am bound tight:
thus, I am
FREE TO SELECT ANY
BASIC ASSUMPTIONS I PLEASE
EXCEPT
that they
MUST NOT CONTRADICT EACH OTHER.
In this way,
I can develop
all kinds of
systems of thought.
I find that
a great many of them,
often to my own surprise,
actually find applications
in the physical world.
And, from past history,
I feel that
many more of them will find
applications in the future.
But I am really driven by
a great curiosity and delight in

the truly remarkable worlds
I create.
They may sound fantastic to
the uninitiated,
but I find them not only
fascinating but
they also throw so much light on
"Just what is human thinking anyway?"

Note that he has a very clear
realization of
where freedom ends and
license begins: *
he knows full well that
he cannot introduce anything
into a system which would
destroy the system itself
by contradiction.

Page those pseudo-liberals
who try to introduce into
a system of Democracy
ideas which would
destroy Democracy itself.
They ought to be
OBLIGED TO STATE EXPLICITLY

* See that delightful book by
  C. J. Keyser
  called: "Mathematical Philosophy,
  A Study of Fate and Freedom" (Dutton).

JUST WHAT THEY CONSIDER TO BE
THE FUNDAMENTAL POSTULATES FOR
DEMOCRACY;
they would probably find that
a good many things they now advocate
CONTRADICT
some of their fundamental ideas.
And they would be obliged to admit
that
even FREEDOM OF SPEECH itself
is necessarily limited,
since it must not be used
to contradict
the other postulates for Democracy.
So that even in a Democracy
it is NOT LOGICAL to
allow an enemy of Democracy to use
Freedom of Speech
to destroy Democracy!

Similarly
FREEDOM OF ENTERPRISE
must also necessarily be limited
by the other postulates of Democracy.

And so you see how
Mathematics can throw light
on various subjects
which many people discuss
glibly and carelessly

since they have never been trained
to examine ideas
With that METICULOUS CARE
with which a mathematician works.
THERE is a model for straight thinking
which we all MUST try to imitate.
This is not the
noisy argumentation of
the pseudo-thinkers.
Rather it is
quiet,
honest,
careful,
COMPETENT.

The Moral: Do not be NAÏVE—
use the methods of
Mathematics.

XV. PRIDE AND PREJUDICE

We hope that by this time
you no longer feel that
"There is only one Geometry—
 good old Euclid;
 he may have given me
 many a headache
 in high school,
 but at least he is respectable."
And that you will not be inclined
to agree with Mrs. Hardy in
"Andy Hardy Meets Debutante"
when she says:
"Nice people never change";
but that you are prepared to agree
with the late
Supreme Court Justice
Benjamin N. Cardozo
who said:
"We are to beware of the
 insularity of mind that perceives

153

in every inroad upon habit
a catastrophic revolution."

But, if you are still in doubt,
we want to give you,
in this chapter,
a simple but most charming
illustration of a Geometry which
will limber up your mind
so beautifully that
you will be prepared to
glide through this changing world
with ease.

We must however call to
your attention
the fact that
whereas
new ideas are many and welcome
in Mathematics,
still
they are not just the
ravings of any "radical" child.

With this brief reminder,
let's go.

You know, of course, that
in Euclidean Geometry,
a plane, or even a line,

154

has an infinite number of points.
Now, in the Geometry which
we shall presently describe,
this is not so;
here
there are only
TWENTY–FIVE POINTS
in the entire Geometry;
and it is therefore called
A FINITE GEOMETRY.

Let us designate the
25 points by the
25 letters of
the English alphabet,
from A to Y inclusive.
And let us arrange these letters
in three blocks as follows:

A	B	C	D	E		A	I	L	T	W		A	X	Q	O	H
F	G	H	I	J		S	V	E	H	K		R	K	I	B	Y
K	L	M	N	O		G	O	R	U	D		J	C	U	S	L
P	Q	R	S	T		Y	C	F	N	Q		V	T	M	F	D
U	V	W	X	Y		M	P	X	B	J		N	G	E	W	P

Now let us make the
following
ASSUMPTIONS:
(1) A "straight line" shall mean
    any row or column in
    any of the three blocks above.

(2) A point-pair shall be
considered "congruent" to
another point-pair when
both pairs occur in rows
(or both in columns),
AND IF
the number of steps
between the points
is the same in both pairs.

A	B	C	D	E		A	I	L	T	W		A	X	Q	O	H
F	G	H	I	J		S	V	E	H	K		R	K	I	B	Y
K	L	M	N	O		G	O	R	U	D		J	C	U	S	L
P	Q	R	S	T		Y	C	F	N	Q		V	T	M	F	D
U	V	W	X	Y		M	P	X	B	J		N	G	E	W	P

Thus,
AB is congruent to HI;
QS is congruent to MX
(even though QS is in a row of
the FIRST block,
whereas MX is in a row in
the SECOND block);
AK is congruent to WD;
etc.
But
AB is NOT congruent to GI,
etc.

Note also that

156

AB is congruent to TP
because we shall consider
the number of steps from T to P
(in the first block)
to be one:
for, when we come to
the end of a row (or column)
we continue to count forward
by jumping to the beginning of
the same row (or column).

Note also that
AB is NOT congruent to AF
since AB is in a row,
and AF is in a column,
whereas in (2) on page 156
it was said that
both must be in rows,
or both in columns,
but not
one point-pair in a row and
the other in a column.
Note also that the word
"congruence" here
does NOT mean the same thing as
in Euclidean Geometry,
where "congruence" involves
"distance,"
and two line-segments are
congruent only if they

can be made to fit throughout;
whereas here,
there is no question of
"distance" or "fitting,"
but merely of
"number of steps."

Also,
"straight line" here does not have
the same significance as in
Euclidean Geometry—
since here it means only
any row or any column.
It would be better,
in order to emphasize
these distinctions,
to arrange the blocks
as follows:

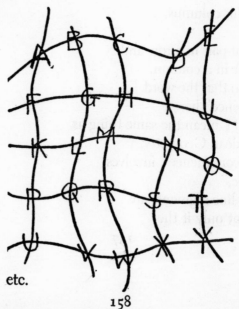

etc.

And, surely, now
you are not in the least
disturbed by
having *ABCDE* called
a "straight line."
Why?
Because this is according to
the "rules of our game":
see (1) on page 155.

Let us now say that
two straight lines are
"parallel" if they have
no point in common.
This is a pretty good use of
the word "parallel,"
is it not?
Hence,
KLMNO is parallel to *FGHIJ*,
since none of the letters in *KLMNO*
occurs also in *FGHIJ*;
but
ABCDE is NOT parallel to *BGLQV*
because they both have
the point *B* in common.

Here of course
there is no question of
two lines meeting if
"sufficiently prolonged,"

because here
no prolonging is possible,
since there are no points beyond
those we have exhibited;
the entire set of points
is visible at a glance.

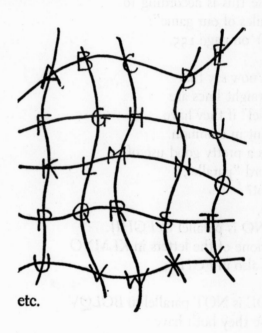

etc.

Note that
whereas in Euclidean Geometry
two lines which are "parallel"
not only have no point in common,
but also

160

they are
"everywhere equally distant,"
but here,
where "distance" has no significance,
this second property disappears,
so that two such lines as
BGLQV and *EJOTY*
may be considered parallel
without worrying us at all.

In other words,
one way in which
the mathematician is enabled
to make up a new system
is
to take some
old familiar word,
like "parallel,"
examine into its various properties,
retain some of these but
discard others,
thus obtaining
a new freedom
without entirely
cutting loose from the past.

There may be a moral here for T.C.—
a hint of
how to proceed when
looking for something new:

Not to break entirely with the past,
but to mold it and modify it
to suit new needs.
Remember that
an entirely new Geometry was found
by merely changing
ONE POSTULATE (see page 143) !

Let us now see
what a triangle looks like
in this new setup.
Take for instance
the triplet of points, H, L, and R;
they form a triangle
whose vertices are of course H, L, R;
and whose sides are
HL, LR, and HR.
And since straight-line-segments
are taken only along
rows or columns,
but not diagonally,
we find the side HL
in the THIRD block on page 156,
LR in the SECOND block,
and HR in the FIRST block.
Thus the triangle here
is completely dismembered,
like the lady in
Picasso's "L'Arlésienne."
Incidentally

162

it so happens that
HL, LR and HR are all congruent
(since they are all in columns
and each is a two-step pair).
So that our triangle is equilateral;
whereas triangle ABJ is isosceles
but not equilateral,
and triangle AST is neither.

A circle here is defined, as usual,
as a set of points such that
any one of them taken with the center
gives a point-pair which is
congruent to every other such pair:
thus,
if we take A as center,
and AB as radius,
then points B, E, I, W, X, and H
all lie on the circle,
because AB, AE, AI, AW, AX, and AH
are all congruent;
so that here
a circle has only six points
on its circumference.

A	B	C	D	E		A	I	L	T	W		A	X	Q	O	H
F	G	H	I	J		S	V	E	H	K		R	K	I	B	Y
K	L	M	N	O		G	O	R	U	D		J	C	U	S	L
P	Q	R	S	T		Y	C	F	N	Q		V	T	M	F	D
U	V	W	X	Y		M	P	X	B	J		N	G	E	W	P

Now you would be surprised to find
that nearly all
the Euclidean postulates
are meaningful here,
in spite of the
meagerness of this little Geometry;
and a great many of
the Euclidean theorems
hold here also.
For example,
the three altitudes of a triangle
are concurrent;
so are the three
perpendicular bisectors of the sides;
and also the three medians.
Furthermore
the point of concurrence of the medians
lies on the line joining
the other two concurrency points
mentioned above,
and divides it in the ratio 2 : 1
just as in Euclidean Geometry.

Here also:
If two sides of a quadrilateral
are congruent and parallel,
so are the other two sides.
The diagonals of a parallelogram
bisect each other.
The diagonals of a rhombus

164

are perpendicular.
At each point of a circle,
there is one and only one "tangent"
(that is,
only one line which has
a single point in common with
the circle) .

In fact
a whole theory of conic sections
is possible here,
etc., etc.

Do we hear some cynic say again:
"So what?"
If so,
let us point out:
(1) This Finite Geometry actually
    arises in connection with
    certain problems in
    Algebra and Number Theory!
    (You see, Mr. Cynic,
    do not be in a hurry
    to call a thing useless—
    perhaps you think it is useless
    only because your knowledge
    is limited!)
(2) Note that
    this entire subject is built up
    by logic alone,

without diagrams—
thus, a triangle,
as we said before (see page 162)
no longer looks like a triangle;
circles look very STRANGE
(completely dismembered!),
and so on.
And all this HELPS to
EMPHASIZE relationships
without OLD PREJUDICES!
And to realize that
Geometry is really a matter of
LOGIC and NOT of DIAGRAMS!

The Moral: Beware of
superficial appearances!
Get behind them
with a clear head,
and find out
what is back of that
good old propaganda.
This process may lead you
to some strange,
"dismembered,"
modernistic things;
but do not let the
strangeness scare you;
DEEP–SET PREJUDICE
MAY BE WORSE THAN
STRANGENESS!

XVI. TWICE TWO IS NOT FOUR!

Perhaps you have now become
so modernistic that
you are really no longer disturbed
by the funny, dismembered
triangles and circles,
and that you are even
willing to grant that
there is some advantage in all this.
But
"Twice two is not four"!!
This is really too much.

Let us try to make this clear.
As we have seen,
one way of making progress in
Geometry
is to take some old familiar word,
like "parallel,"
and limber up its meaning
just enough to make it possible to
put it to some new use
(see page 161).

Now, perhaps you do not realize it
but
you have already had
similar experiences in Algebra:
For instance,
when you began the study of Algebra
in your first year in high school,
you became acquainted with
"negative" numbers,
as distinguished from
the ordinary, "positive" numbers of
Arithmetic.

Thus, if we represent numbers by
points on a line,
we may place the positive numbers
to the right of zero, thus:

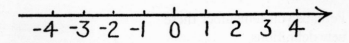

and the new negative numbers
to the left.
So that in Algebra
you were introduced to
a whole new set of numbers which
do not come into Arithmetic at all;
and, as you doubtless know,
these new numbers are just as

"practical"
as the old ones,
since, for example,
a temperature of 5° below zero (— 5°)
is just as "real" as
a temperature of 5° above zero—
ask Admiral Byrd!
And a debt of 50 dollars (— $50)
is just as "real," although not as pleasant, as
having 50 dollars in your pocket—
ask any man who is being sued for
debt!

And you remember you then had to
find out
what is to be meant by
"adding" these new numbers,
and "multiplying" them.

Perhaps you remember these
new definitions,
and perhaps you remember also
how strange they seemed to you
at the time.
Thus the rule for "adding"
a positive number and
a negative number is:
"Take the DIFFERENCE of the numbers
 and then prefix to the result
 the sign of the larger one."

Or, in plain English,
if you wish to "add"
—5 dollars and 3 dollars,
the answer is —2 dollars,
because
if you OWE 5 dollars (—$5)
and HAVE 3 dollars ($3)
and wish to
BALANCE YOUR ACCOUNT,
you pay part of your debt and
still OWE 2 dollars.
In other words,
"addition" now means
"balancing accounts."
And you are not in the least
disturbed by the fact that
when you "add," you sometimes
really "subtract,"
as you may have heard some
youngsters say in high school!
The fact is that they are using
"add" in the algebraic sense,
and "subtract" in the
arithmetic sense;
but there is really no confusion
if you realize clearly that
"add" in Algebra does not have
the same meaning as
"add" in Arithmetic.

Thus,
as we said before,
you have already had some experience
in changing the meaning of
a word in Mathematics
in order to have
progress.

And those of you who have had
some elementary Physics
are familiar with
still another meaning of the word
"add":
For instance,
suppose a force of 2 pounds
acts on a body, A,
in the direction indicated;
and suppose another force of 2 pounds
acts on A in another direction,
as shown:

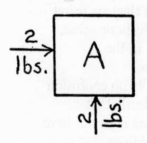

The question is
in which direction will A

actually move,
and how great a force is
actually pushing it?

You probably remember that
this problem is solved by
what is known as
"The parallelogram of forces,"
as follows:
Draw two lines, *BD* and *BE*,
representing the two given forces
in magnitude and direction,
thus:

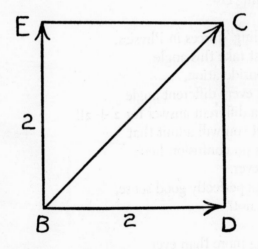

Then complete the parallelogram by
drawing *DC* and *EC*;
then the line *BC* is

173

the "resultant" or "SUM" of
the two given forces;
and,
from triangle BDC it is quite easy
to calculate the length of BC,
which in this case is about
        2.83 lbs.
So that here
2 + 2 is NOT 4 but 2.83!
And,
if the angle at B did not
happen to be a right angle,
the sum of 2 and 2 would be
something else.
Thus,
in "adding" forces in Physics,
we must take this angle
into consideration,
and for every different angle
we get a different answer for 2 + 2!!
And yet you will admit that
there is no confusion here
whatsoever.
It makes perfectly good sense,
does it not?
Now,
realizing more than ever
that in Mathematics we are free
to make any assumptions which
we find useful,
so long as they

DO NOT CONTRADICT EACH OTHER,
we see that in this way
many different Algebras,
as well as different Geometries,
can be constructed—
and actually have been.
Some of these have found
marvelous applications,
as, for example,
Boolean * Algebra (used in Logic),
by means of which
you can test, for instance,
the consistency of
a whole set of legal statements
by expressing them in
"algebraic" form,
in Boolean Algebra,
and applying to these expressions
the rules of manipulation of
this Algebra!
The implications of this tool
as an aid to clear thinking
for general "life situations"
are not yet fully appreciated! **
And now to give you an idea of
an Algebra entirely different from
the one to which you are accustomed,

* Boole first started this idea (1850).
** See papers on Logic by
   Alonzo Church of Princeton University
   (Galois Institute Press, 1942).

we shall give you here
a little FINITE ALGEBRA
constructed by
Professor Emeritus E. V. Huntington of
Harvard University.*
This little Algebra has as its
basic POSTULATES
nearly all the postulates of
ordinary Algebra,
EXCEPT ONE;
and yet
it has only NINE numbers in it:

0, 1, 2, 3, 4, 5, 6, 7, 8.

But,
you must not think of these as
being the ordinary numbers
which you know;
think of them rather as
nine SYMBOLS
which you are to manipulate
according to certain rules—
as if you were learning to play
some new parlor game.

Thus we shall give you
two tables in which you may
look up
the "sum" or "product" of
any two of the numbers:

* "The Fundamental Propositions
  of Algebra" (Galois Institute Press, 1941).

176

SUM TABLE

	0	1	2	3	4	5	6	7	8
0	0	1	2	3	4	5	6	7	8
1	1	2	0	4	5	3	7	8	6
2	2	0	1	5	3	4	8	6	7
3	3	4	5	6	7	8	0	1	2
4	4	5	3	7	8	6	1	2	0
5	5	3	4	8	6	7	2	0	1
6	6	7	8	0	1	2	3	4	5
7	7	8	6	1	2	0	4	5	3
8	8	6	7	2	0	1	5	3	4

PRODUCT TABLE

	0	1	2	3	4	5	6	7	8
0	0	0	0	0	0	0	0	0	0
1	0	1	2	3	4	5	6	7	8
2	0	2	1	6	8	7	3	5	4
3	0	3	6	4	7	1	8	2	5
4	0	4	8	7	2	3	5	6	1
5	0	5	7	1	3	8	2	4	6
6	0	6	3	8	5	2	4	1	7
7	0	7	5	2	6	4	1	8	3
8	0	8	4	5	1	6	7	3	2

Here we have

$$2 + 2 = 1$$
$$7 + 1 = 8$$
etc.

And

$$5 \times 7 = 4$$
$$2 \times 2 = 1$$
$$8 \times 0 = 0$$
etc.

177

What interests us here is
the fundamental idea that
various Algebras,
like various Geometries,
are possible;
that twice two
may be four or not four,
depending upon
the Algebra in question;
that all these
Algebras and Geometries are
MAN–MADE;
that, therefore,
there is nothing ABSOLUTE
about any one of them;
that none of them represents
THE truth;
and yet
all or many of them are
extremely useful;
that Man,
even though he has not found,
and probably never will find,
THE truth,
yet
he can, by his ability to
THINK,
do very well for himself,
if he would only use it!

This does not mean that
Man can say to God:
"See, I am as good as You are.
  I really do not need You at all.
  I can get on very well with
  my own REASONING POWER."

Not at all!

We maintain, on the contrary,
that
Man is so far from being
as good as God
that he will probably NEVER know
THE truth:

this emphasizes
Man's HUMILITY rather than
his ARROGANCE!

Thus
since Man has only
his OWN REASON,
and NOT God's,
let him use it
to the best of his ability,
and he will get some
very respectable results,
but let him never brag that
he "knows" THE truth!

181

The Moral:　At the end of Chapter VII
of Part I
we said:
"Be a man—not a mouse!"
And now we add
"Be a man—but
do not try to play God!"
In short, T. C.,
"BE YOURSELF!"

XVII. ABSTRACTION—MODERN STYLE

You remember that
we pointed out on page 82
the importance of ABSTRACTION,
and we promised there
to say more about
ABSTRACTION as practiced by
the MODERNS.

Well,
now that you have seen
some strange new
Algebras and Geometries,
you are prepared to enjoy
a more abstract system:
Here, instead of having
points or numbers as our
ELEMENTS,
we shall take the four "objects":
"chalk," "red," "chair," "desk."
The sum and product of
any two of these elements
may be obtained from

the following two
tables:

	Chalk	Red	Chair	Desk
Chalk	Chalk	Red	Chair	Desk
Red	Red	Chair	Desk	Chalk
Chair	Chair	Desk	Chalk	Red
Desk	Desk	Chalk	Red	Chair

MULTIPLICATION TABLE

	Chalk	Red	Chair	Desk
Chalk	Chalk	Chalk	Chalk	Chalk
Red	Chalk	Red	Chair	Desk
Chair	Chalk	Chair	Chalk	Chair
Desk	Chalk	Desk	Chair	Red

Thus Chair + Red = Desk
and Red × Chalk = Chalk
etc., etc.

Of course the words
chalk, red, add, multiply, etc.,
do not have the ordinary meanings,
but are manipulated in accordance

184

with whatever rules or postulates
we have chosen.

In order not to confuse the reader,
abstract addition and multiplication
are sometimes designated by
$$\oplus \text{ and } \otimes$$
to distinguish them from
ORDINARY addition and multiplication,
for which we use
$$+ \text{ and } \times$$
without any circles around them.
Of course $\oplus$ and $\otimes$ do not have
a SPECIFIC meaning,
and therefore give to
the modern mathematician
greater FREEDOM to
INVENT and DEVELOP
all kinds of systems.
And some of these systems
may turn out to be of great
practical value if and when
they are applied to
definite problems.

But the mathematician's job is
to have them ready
in abstract form so that
they may be used in any situation
wherever they may be helpful.

And so you see that
an important modern trend in
Mathematics
is toward more and more
ABSTRACTION.
And you can see what a
source of
FREEDOM and POWER it is.

In Chapter XX
you will see how
this same trend toward
abstraction
vitalizes and enriches
Modern Art as well.

The Moral: Go MODERN:
            learn to
            APPRECIATE THE ABSTRACT.

XVIII. THE FOURTH DIMENSION

"Very well," you will say,
"I am quite willing to grant now
  that there may be
  various Geometries and
  various Algebras
  provided we start with
  various assumptions and
  give some new meanings to old words.
But I still feel that
there is such a thing as
THE truth:
I admit that
'Twice two is four' is
NOT such a good sample of it
as I once thought.
But how about
the actual physical world?
Surely here we are not so free
to make any assumptions we please.
Here we must be governed not only
by clear mathematical thinking—
which forbids only one thing:
self-contradiction—

but, in the physical world
we are also held down by
OBJECTIVE FACTS,
which I still regard as
THE truth!"

And, since you are quite an
educated T.C.,
you may proceed as follows:
"Suppose, for instance, that
   someone,
   let us call him Mr. K,
   wishes to measure
   the distance from O to A:

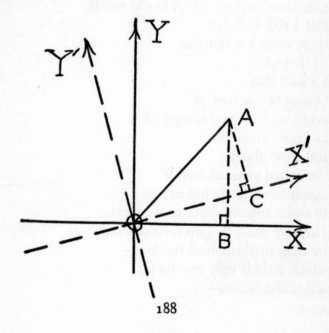

and suppose that, for some reason,
he cannot measure it directly,
but is obliged to do it indirectly
(as in so many problems in
Trigonometry),
by measuring OB and AB instead,
using the axes X and Y.
He can now calculate OA by using
the Pythagorean Theorem:

$$OA = \sqrt{(OB)^2 + (AB)^2}.$$

Further,
suppose another observer, Mr. K',
finds it convenient to use instead
the axes X' and Y'.
He then measures OC and AC
(instead of OB and AB as Mr. K did)
and calculates OA by means of

$$OA = \sqrt{(OC)^2 + (AC)^2}.$$

But note well that
although K and K' have made
DIFFERENT MEASUREMENTS,
yet they get the
SAME ANSWER!
And," continues T.C. who
has really been impressed by
Part I and sees
the possibilities,
"and, mind you,
if a still more
individualistic gentleman,
Mr. K''',

prefers to use the axes X″ and Y″
shown below:

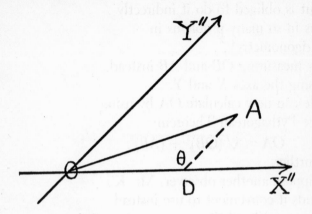

he can still calculate OA,
(although he now measures OD and AD)
by using a well-known formula from
Trigonometry:
$$AO = \sqrt{(OD)^2 + (AD)^2 - 2\,(OD)\,(AD)\cos\theta},$$
and AGAIN gets the SAME ANSWER!
Thus I conclude that
in spite of the individualism of
K and K′ and K″ and others,
still they can all
'do business' together
because they agree on the result,
namely, in this case,
the length of OA,
which I therefore believe to be
an objective fact.

And it is for this reason that
I believe in the possibility of
The Good-Neighbor Policy,
in which various people can have
a certain amount of individualism
and yet can AGREE
on certain FACTS."

Well, T.C.,
we agree with nearly everything
you said,
but
we still maintain that
the human race does not,
and probably cannot,
"know the facts."
And yet
your idea of
The Good-Neighbor Policy
is still acceptable:
In order to show clearly
what our position is,
we must make a little detour
to discuss what is known as
"Dimensionality."

As you know so well,
a point in a plane may be
designated by a pair of numbers:

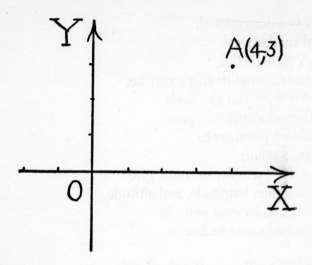

A(4,3)

Thus point A is designated by (4,3)
because it is reached by going
four units to the right of O
and three units up.
And similarly
for any other point in the plane.
We therefore say that
a plane is a "two-dimensional space."
Note also that
it takes 2 numbers to locate
a point on the surface of a globe,
namely,
its latitude and longitude;
and, therefore,
the SURFACE of a globe is

also two-dimensional.
And similarly for
ANY SURFACE,
no matter what its shape may be.
And now, as you also know,
in three-dimensional space
it takes 3 numbers to
locate a point;
thus, for example, we must give
the latitude, longitude, and altitude,
in order to locate a point in
the actual world we live in.

We must call your attention here
to an important idea:
In the above discussion we have been
talking about locating
a "point";
but suppose we choose some other
"element,"
instead of "point,"
say, "circle,"
and imagine any space which
we examine
as filled with circles of
different sizes,
with various centers.
If we now wish to discuss
the "dimensionality" of
the space in question,

we should have to proceed
as follows:
Take first an ordinary
Euclidean plane,

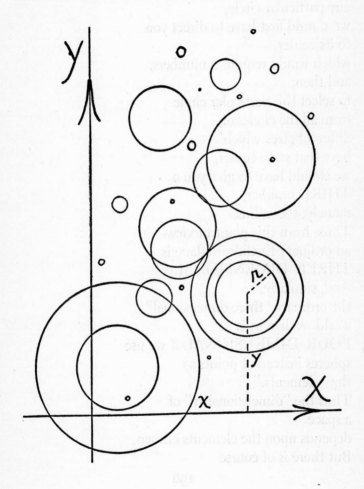

and imagine it to be covered with
circles instead of points,
as mentioned above;
now, to locate
any particular circle,
we should first have to direct you
to its center,
which would require 2 numbers,
and then,
to select this particular circle
from all the circles of
different sizes which
have that same center,
we should have to give you a
THIRD number,
namely, the radius.
Thus, from this point of view,
an ordinary Euclidean plane is
THREE–DIMENSIONAL!
And, similarly,
the ordinary "three-dimensional"
world we live in is
FOUR–DIMENSIONAL if we use
spheres instead of points as
the "elements."
Thus the "dimensionality" of
a space
depends upon the elements chosen.
But there is of course

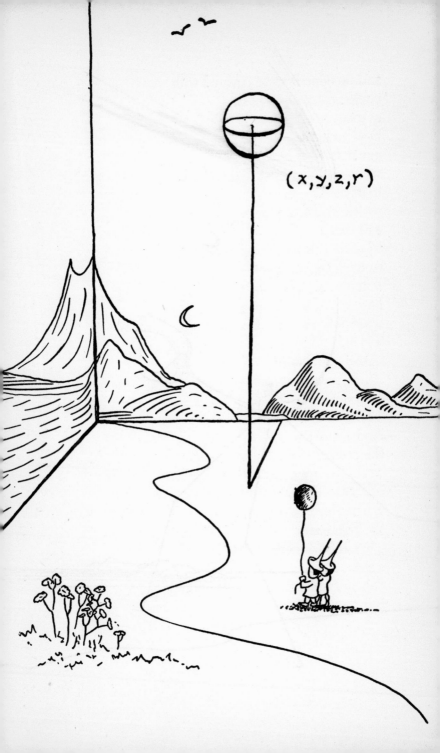

(x, y, z, r)

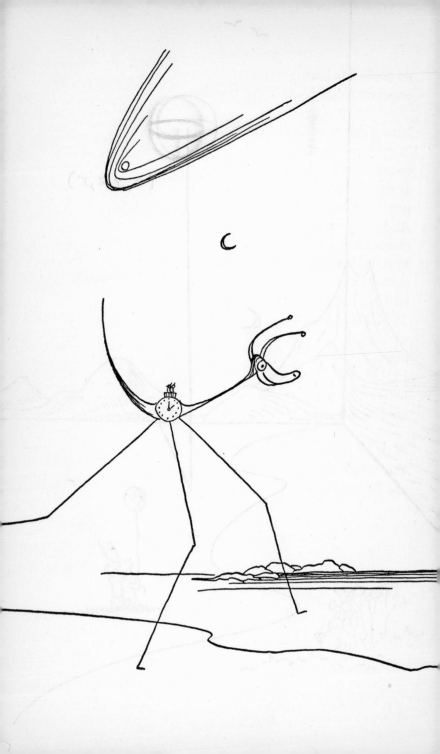

no confusion here,
provided we SPECIFY the elements.

Now it has been found convenient
in modern Physics
to use "events" instead of "points"
as the elements in describing
physical phenomena.
And since every event is
characterized by
FOUR numbers,
namely, the
latitude, longitude, altitude, and
TIME of its occurrence,
we may therefore speak of
living in a
FOUR–DIMENSIONAL WORLD
without being either
confused or mystical!

Now let us see what bearing
this has on the discussion at
the beginning of this chapter.

XIX. PREPAREDNESS

As we have already seen
(see page 189),
the length of a line on
a Euclidean plane,
using any rectangular axes,
is given by the formula
$$d = \sqrt{x^2 + y^2}.$$
And, similarly,
in three-dimensional
Euclidean space
$$d = \sqrt{x^2 + y^2 + z^2}$$
where $x = OB$, $y = BC$, $z = AC$
(see the diagram on page 201).
Note that in this
three-dimensional space,
$x^2 + y^2$ taken alone,
without the z,
is no longer the same for
different sets of
rectangular axes.
Thus, if instead of
changing the axes, we move OA
to a new position,

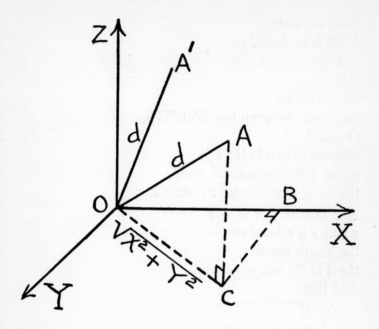

such as OA',
we see that
$\sqrt{x^2 + y^2}$ (or $\sqrt{(x')^2 + (y')^2}$)
is merely the
"projection" or "shadow"
of OA (or OA'),
and of course
the SHADOW of an object
may CHANGE in length as
the object is moved around,
without changing the
length of the object itself.

Now similarly
it has been found in
modern Physics
that
if you take the
"interval" between two EVENTS,
O and A
(instead of two POINTS)
in our four-dimensional world,
the thing that remains constant is

$$\sqrt{x^2 + y^2 + z^2 + \tau^2}$$

where τ is related to
the fourth number,
the TIME (see page 199).
And that

$$\sqrt{x^2 + y^2 + z^2}$$

is no longer a constant,
just as
$\sqrt{x^2 + y^2}$ did not remain constant
when going from 2 to 3 dimensions
(see page 200).

Translated into plain English
this says that
two observers, K and K',
who are moving
relatively to an object with
different but uniform velocities,
do NOT both get the same result for
the LENGTH of that object:

that the length of an object is,
you might say,
just a three-dimensional
"projection" or "shadow"
of a four-dimensional "interval"
(see pages 201 and 202).*
"Well," replies T.C.,
"you are beginning to
   sound a little mystical,
   and yet
when I look at the formulas
you gave me,
I follow you all right.
But, after all,
to go back to the
question raised on page 192,
all you have really done is
merely to say that
in modern Physics
you no longer regard
the length of an object as
an immutable fact,
independent of the particular observer
(as it was regarded before Einstein),

---

* If you want to know more about
   this interesting point,
   see
   "The Einstein Theory of Relativity,
   the Special Theory"
   by Hugh Gray and Lillian R. Lieber
   (Galois Institute Press).

but, anyway,
your four-dimensional 'interval'
$$\sqrt{x^2 + y^2 + z^2 + \tau^2}$$
IS THE SAME
for K and K′,
so THIS is now the
OBJECTIVE FACT
instead of
$$\sqrt{x^2 + y^2 + z^2}.$$
The principle, however,
is still the same:
there ARE physical facts,
and we are gradually
finding them out."

We see, T.C.,
that you are intelligent but
nineteenth-century minded;
for,
the modern twentieth-century
physicist
realizes now that
ANY "facts" that he finds
are tentative.
They represent the best we have
in the light of
all known observations and
experiments,
at a given time.

But he sees clearly that
even these observations
are Man's observations,
subject to the limitations of
his senses and
his mind,
and should in no way be regarded
as "true."
As Einstein says:
"*Alles was wir machen ist falsch.*"
"Everything we make is false."

What then is the use of it?

The obvious answer is:
"The proof of the pudding is
  in the eating!"
That is,
if our Science enables us
to get around more easily in
this complicated world,
even if it is no more than
a mere system of "bookkeeping"
or a mere "mnemonic,"
it serves to correlate our
various observations
so that we can at least
remember them and
envisage them better.

In short,
the modern physicist,
making his observations
in his human way,
finds it convenient to take as
postulates for Physics
certain things which he
repeatedly observes,
and then
he develops by means of
his human logic
certain consequences of
these postulates.
And, finally,
he makes more observations in order
to see whether
he will OBSERVE these consequences
which he arrived at by his logic.
If he does,
he calls his theory good,
but he is under no delusion that
the theory will remain good
in the light of
ALL future observations.
Naturally so long as he does
observe what he predicted,
he feels good,
and we cannot begrudge him
this feeling of satisfaction.
For it is only fair to admit that

if we compare
a scientist's predictions,
say Einstein's,
with those of other people,
we cannot help but be
IMPRESSED with
the much greater success which
he has.
Thus when Einstein predicted in 1916
that
if on such and such a date (1919)
you should go to such and such a place (Africa)
and set up your camera
and take a picture of the stars,
you would find that their positions
on the photographic plate
had shifted from their
normal positions
by a certain tiny amount
(about 1.75 seconds of arc).
And when the scientists followed
his directions,
they found just what he predicted!

If anyone outside of Science
can match this power of prediction
we shall admit that
he has as good an approach as
the scientific one!
But we need hardly say that

this power of prediction
has NOT been matched
by the average
science-heckling,
loose-thinking
loud-speaker!
And that is why
we claim that
scientific predictions are
a triumph of
CLEAR THINKING
even if they are not
THE ABSOLUTE TRUTH.

And so,
a modern scientist no longer speaks of
"objective facts,"
but of
"invariants under transformations."
Thus

$$\sqrt{x^2 + y^2}$$

is an INVARIANT under a
ROTATION OF AXES in a
TWO-dimensional Euclidean space,
but it is
NOT an invariant under
such a transformation of axes in
THREE-dimensional Euclidean space
(see page 202).
And you will no doubt agree

that this is a
more precise
as well as a more modest
way of speaking.

And,
in this way,
the scientist also holds himself
in readiness for change!
For if new observations are made,
or if he reconsiders his
basic ideas,
thus requiring the introduction of
new transformations,
he will expect to give up
his old invariants for new ones.
And, being prepared for
this possibility,
he will not be as startled by
changes
as were the 19th century physicists
when Einstein introduced
his new system.
So that now,
not only has this new system
been accepted because
it is more adequate than
the old one,
but the physicists have,
as a result of all this,

a much more wholesome outlook
on their entire activity.

The Moral: The modern viewpoint
            demands
            greater flexibility of
            mind and
            preparedness for change
            Pull your mind out of
            those muddy old ruts!
            And adapt yourself to
            a continually
            CHANGING world.

XX. THESE MODERNS

Let us now return to
the top floor of the Totem Pole,
where twice two
may or may not be four,
where triangles,
as well as ladies,
are dismembered—
in short,
where you find
the MOST MODERN in
Mathematics, Art, Music, etc.

And let us see what
these DIFFERENT domains have
IN COMMON,
that makes them all MODERN,
though they seem so unlike each other
on the surface.

The modern trends seem to be:

(1) Man has begun to realize that

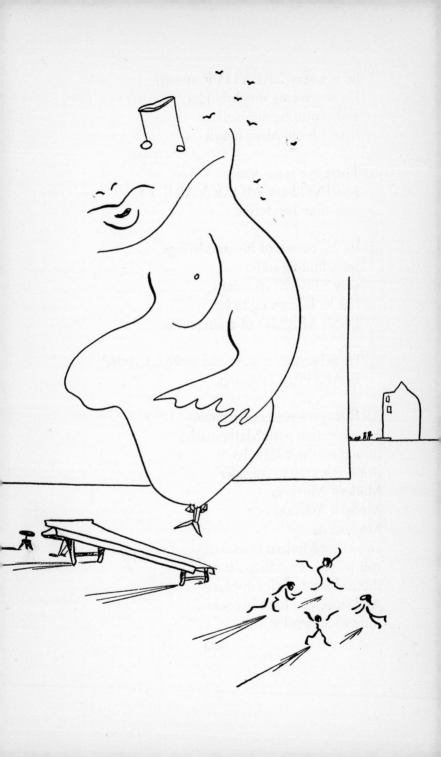

he is a very CREATIVE animal.
He is growing much bolder
and venturing out further
from his old playgrounds.

(2) There is consequently
INFINITELY MORE VARIETY
now than heretofore.

(3) In the course of his wanderings,
he is finding some
very STRANGE things,
but he is learning to be
LESS AFRAID of strangeness.

(4) He is becoming more and more interested
in ABSTRACT things.

All this you have been realizing
in connection with Mathematics
throughout this little book.
But if you stop to consider
Modern Music or
Modern Aviation or
Modern Art or
any other Modern domain,
you will see that there, too,
these characteristics appear.
And since you have become
somewhat familiar with

strange new things,
and have grown to like them
(we hope),
you will doubtless look
with wonder and admiration
at the drawings in this chapter.
For a MODERN EDUCATION must include
at least a little familiarity
with the moderns in various domains.

You may expect that
these drawings
will be strange and dismembered,
but you now realize that
strangeness is a characteristic
of modernism—
even in Mathematics and Physics.

And you will not be surprised
to find that
modern art has
INFINITELY MORE VARIETY
than old-fashioned art,
just as Mathematics today has
INFINITELY MORE VARIETY
than it used to have.

And, above all,
bear in mind the admonition
we gave you on page 71,

when we asked you
not to inquire of
ANY top-floor man,
whether mathematician or artist,
"What is the practical use of
   what you are doing?"
"What does this mean for
   the Average Man?"
For, as we told you,
nobody knows!
The products of the top floor
are natural phenomena,
the most interesting of
all human documents—
and if they ever
come back as a
first-floor gadget,
this is NOT
the most interesting thing
about them!

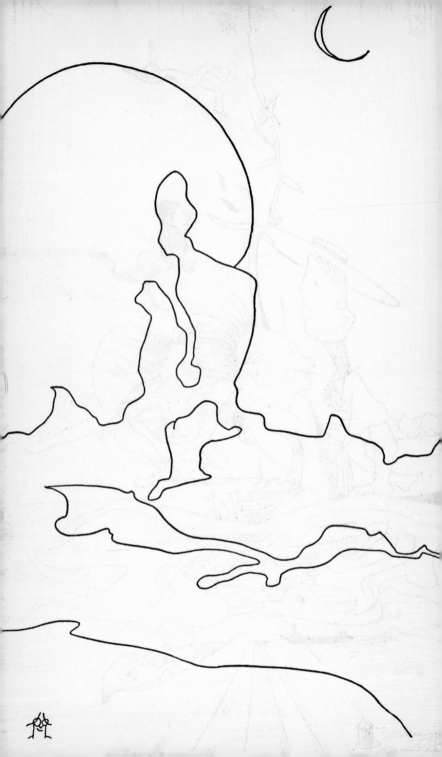

THE MORAL

It is possible to have
agreement
and yet permit
different viewpoints
(Chapters XVIII and XIX).

Unless we compare
the different viewpoints
we cannot even speak of
invariants (page 208) —
which makes
isolationism and provincialism
ridiculous,
and tolerance essential.

These invariants may be derived
by various observers
"With equal rights and equal success" *
(page 190).

* See "Einstein's Theory of Relativity" by H. G. and
L. R. Lieber (Galois Institute Press).

But what is it that
each observer has
the RIGHT to do?
Obviously ONLY to do his BEST
(judged by strict standards) :
To measure as accurately as
possible
(as judged by the best
laboratory practice) ;
to think straight
(as judged by the best standards of
modern mathematicians and logicians),
and NOT MERELY TO HECKLE!

Modesty and humility
and self-reliance (page 181)
should characterize man's activity.

Since his knowledge is
only tentative (page 204)
he must be
PREPARED FOR CHANGE (page 210).

But he must progress with
a minimum of upheaval (page 162),
respecting tradition without
being a slave to it (page 117).

Clear thinking combined with
careful observation are his
most "practical" weapon (page 206).

"Common sense" can be
enlarged and developed
and should not remain
childish (page 93).

"Human nature" is
NOT synonymous with
"money-grabbing" and "throat-cutting":
Man is a much more
complex and interesting creature
(pages 67 and 214).

War is not to be blamed on Science
(Chapters V and VI).

We can have
freedom without anarchy (page 174).

Democracy is
essential to
human accomplishment
(page 64 and Chapters XVIII and XIX).
But we must be loyal to
its basic principles or
we cannot have it at all
(page 149).

And so on and so on.

No doubt you can find
many more morals of this kind
in Mathematics and Science and Art,
for this little book is
only a small sample of
this point of view
from which we consider
not so much the techniques
(which here are only incidental)
as the general methods of
Man's successful accomplishment.
Perhaps we can learn from them
how to be equally successful
in thinking about
the social sciences, for instance.
For surely,
Man, with so much
ingenuity and originality,
will not let
his social problems
lick him!
BUT THEY WILL NOT SOLVE THEMSELVES!
He must allow his imagination
greater freedom,
as the mathematicians
and scientists
and artists
do;
and, at the same time,
must bear in mind

the limitations of his freedom.

And now please turn back to
the Introduction
and read it again,
and consider it
thoughtfully
in the light of what
you have read in
this little book.
Do you agree with us that
this material really helps to
clarify the meanings of
these concepts?

SUGGESTED READING

E. T. BELL: *The Development of Mathematics* (McGraw-Hill).

G. BOOLE: *The Laws of Thought* (Macmillan).

EINSTEIN and INFELD: *The Evolution of Physics* (Simon and Schuster).

MICHAEL FARADAY: *Experimental Researches in Electricity* (Everyman's Library).

S. I. HAYAKAWA: *Language in Action* (Harcourt, Brace).

L. T. HOGBEN: *Mathematics for the Million* (Norton).

E. V. HUNTINGTON: "The Fundamental Propositions of Algebra" (reprinted by the Galois Institute of Mathematics Press from *Monographs on Modern Mathematics* published by Longmans, Green).

KASNER and NEWMAN: *Mathematics and the Imagination* (Simon and Schuster).

C. J. KEYSER: *The Human Worth of Rigorous Thinking* (Scripta Mathematica Library).
————: *Mathematical Philosophy, a Study of Fate and Freedom* (Dutton).

H. G. and L. R. LIEBER: *The Einstein Theory of Relativity.*
————: *Galois and the Theory of Groups.*
————: *Non-Euclidean Geometry.*
(Galois Institute of Mathematics Press.)

L. L. THURSTONE: *The Vectors of Mind* (University of Chicago Science Series).

J. W. YOUNG: *Fundamental Concepts of Algebra and Geometry* (Macmillan).

Mathematico-Deductive Theory of Rote Learning (Institute of Human Relations, Yale University Press).

Papers on selected topics of modern mathematics (Galois Institute of Mathematics Press).